THE

GREAT

INCEPTION

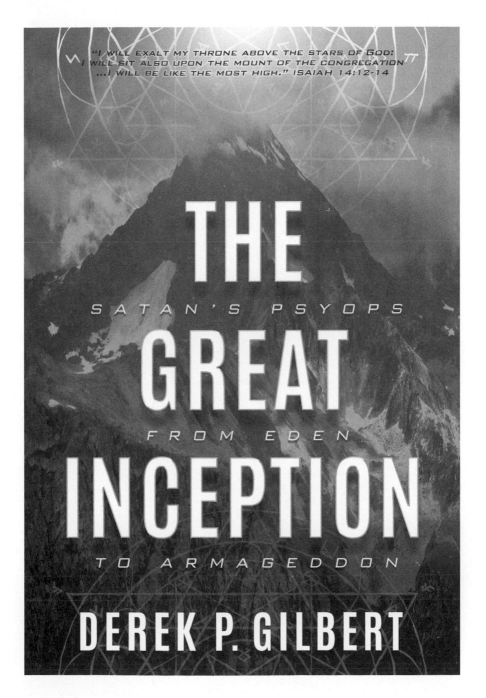

"I WILL EXALT MY THRONE ABOVE THE STARS OF GOD:
I WILL SIT ALSO UPON THE MOUNT OF THE CONGREGATION
...I WILL BE LIKE THE MOST HIGH." ISAIAH 14:12-14

THE
SATAN'S PSYOPS
GREAT
FROM EDEN
INCEPTION
TO ARMAGEDDON

DEREK P. GILBERT

DEFENDER

CRANE, MO

The Great Inception: Satan's PSYOPs from Eden to Armageddon

by Derek P. Gilbert

ISBN: 9780998142647

—◆—

To Sharon,
my wonderful one, who inspires me to be
a better man than I am,
and Nicole,
who makes me more proud than
anything I have ever done.

CONTENTS

Acknowledgments ix

Foreword xiii

Introduction 1

CHAPTER 1: Eden .7

CHAPTER 2: Hermon 31

CHAPTER 3: Babel 51

CHAPTER 4: Sinai 71

CHAPTER 5: Zaphon 101

CHAPTER 6: Carmel 167

CHAPTER 7: Zion 199

CHAPTER 8: Jabal al-Nour 227

CHAPTER 9: What Things May Come 237

CHAPTER 10: The Great Inception 273

CHAPTER 11: Return to the Mountain 287

CHAPTER 12: The Long War 293

Bibliography 301

Notes 311

ACKNOWLEDGMENTS

THIS BOOK WAS inspired by the weekly Bible study Sharon and I produce each week at www.GilbertHouse.org. Without the intellectual stimulus provided by discussing the Word of God regularly with my wife, this book would never have seen the light of day.

Credit is also due the following scholars and men of God for their work, which inspired and helped me focus my thoughts:

1. Dr. Michael S. Heiser, whose work on the divine council and book *The Unseen Realm: Recovering the Supernatural Worldview of the Bible* has been foundational. The book you hold now builds on his work, and if you haven't read *The Unseen Realm* yet, I will remind you to do so a couple more times before you finish this book. (His website is www.DrMSH.com.)

2. Chris White, whose podcasts and books have always pushed me to think and reevaluate what I believe. Please read his books *False Christ: Will the Antichrist Claim to Be the Jewish Messiah?*, *Mystery Babylon: When Jerusalem Embraces the Antichrist*, *Daniel:*

A Commentary, and *The Islamic Antichrist Debunked: A Comprehensive Critique of the Muslim Antichrist Theory*. His website is www.ConspiracyClothes.com.

3. Joel Richardson, whose books *When a Jew Rules the World* and *Mystery Babylon* have been influential, especially on guiding me to realize that Saudi Arabia will play a larger role in the last days than I had previously considered. His website is www.JoelsTrumpet.com.

4. Dr. Timothy R. Furnish, an insightful scholar of Islamic history and eschatology, was sounding an alarm about the danger of violent Mahdism a decade before the Islamic State declared its caliphate. His books *Ten Years' Captivation with the Mahdi's Camps: Essays on Muslim Eschatology, 2005–2015* and *Sects, Lies, and the Caliphate: 10 Years of Observations on Islam* are required reading for anyone who wants to understand where Islam has been and where it is probably going. Tim's website is www.MahdiWatch.org.

To be clear, the conclusions in this book are my own. Please do not assume that these gentlemen agree with all (or even any) of what you're about to read. If I have erred in my analysis, it's on me, not them.

I am indebted to Linda Traylor, a retired English teacher who volunteered to help clean up the final draft of this book. Her contributions were invaluable and are deeply appreciated.

Special thanks to Tom and Nita Horn, dear friends who took Sharon and me on board to partner on a new ministry in 2015, SkyWatchTV. Tom has given us very wide latitude in what we do at SkyWatch, and he is unquestionably the best boss I've ever had. The team assembled under Tom's care is an incredible group of loving and talented people, and the conversations we have around the office make each day intellectually and spiritually invigorating.

Of course, my loving and patient wife Sharon deserves credit for being my constant companion, best friend, and most reliable sounding board.

Get to know her and you'll understand why my dad was convinced his boy had married up.

Finally, all credit for anything I get right in the pages that follow to our Lord and Savior, Jesus Christ. He doesn't need this book to achieve His ends. He doesn't need me blathering into a mic or a camera five days a week. And yet this is the path He's put me on. It is humbling.

May this book be a blessing. And thank you for reading.

FOREWORD

THE TITLE OF this book needs a word of explanation. The word "inception" simply means a beginning, a starting point.

In this context, it implies the onset of something. That is true. It is my belief that we're witnessing the beginning of the endgame of history.

The title, however, was inspired by the 2010 science-fiction film *Inception*, which starred Leonardo DiCaprio as a very special kind of thief—one who could hijack the dreams of others and extract valuable secrets for corporate clients.

The movie's plot focused on a complicated scheme to convince the heir of an energy conglomerate to break up the corporation after his ailing father's death. DiCaprio's character assembled a team to create multiple dream worlds nested inside one another like a Russian doll so the target would be unable to discern reality from a dreamscape and accept the implanted idea, believing it to be advice from a trusted confidant.

The dream worlds created by DiCaprio's character and his team were so realistic that they were taught to create a totem—an item that appears only in the dream state to prevent their conscious minds from being deceived into remaining in an artificial world.

The film raises interesting questions about the nature of reality. Is reality what we experience with our waking minds? If our dreams and visions affect our actions, where do we draw the line between reality and fantasy? And what consequences might there be if someone, or some group, possessed the ability to manipulate the conscious dreams of others?

Those questions are startlingly relevant to the geopolitical situation on planet earth. World events appear to be spiraling out of control. Since the turn of the twenty-first century, the world is a deadlier place. Inflexible religious groups following centuries-old apocalyptic prophecies that promise the establishment of heaven on earth—but only for the faithful—are doing their damnedest to exterminate anyone who stands between them and paradise.

Skeptics and atheists probably read conservative Christians into that statement. That is not who I mean.

The twenty-first century was not two years old when the illusion of invulnerability harbored by Americans was shattered on September 11, 2001. While there are valid questions about what actually happened on that day, an investigation of the actual human actors responsible for 9/11 is beyond the scope of this book.

Besides, we already know the real masterminds: the principalities, powers, thrones, and dominions identified by the apostle Paul, evil intelligences who want to destroy mankind. In the grand scheme, it doesn't really matter if their agents were Saudi or American, Sunni or nominal Christians. The ancient entities behind them employ and discard human agents of all races, colors, and creeds. Throughout history, they have created a false reality—actually, multiple, overlapping, nested, transdimensional false realities—leading generation after generation of human dupes to volunteer to serve in their long war of rebellion against God.

The official story is that the emergence of the Islamic State in the summer of 2014 caught Western governments off guard. This is not at all true. The Islamic State did not simply spring from the sand of Iraq fully formed. As of this writing, more than two years later, we have yet to hear an intelligent explanation for the rise of ISIS from any Western government official. A variety of harebrained explanations have been offered for the appeal and resilience of this murderous group, including a lack of jobs, reaction to unwanted Western interference in their culture, savvy use of social media, and—of course—global warming.

We will set aside the fact that Western governments weren't surprised by the rise of ISIS because their intelligence agencies helped create the monster. That's not a crazy conspiracy theory; thanks to government watchdog Judicial Watch, we know the United States Defense Intelligence Agency warned the Obama administration in August of 2012 that continuing to arm rebels in Syria would lead to an independent Islamic State, destabilizing both Syria and Iraq. But that road leads in a direction that is completely different than the one I have in mind.

I would only note this: Understanding the motives of the Islamic State is simple. Just read their magazine. The leadership of ISIS is very open about it: The caliphate intends to provoke a climactic battle with "Rome" in northern Syria or southern Turkey, at which Isa (their version of Jesus) and the Mahdi (their version of the Savior) will appear and lead Islamic forces to victorious global jihad.

Contrary to the assurances of Western politicians, the apocalyptic worldview of the Islamic State is not a corrupt or hijacked version of Islam. It's a return to Islam's original design, and one that is held by a startling number of Muslims.

Meanwhile, over the last couple of years, messianic declarations from Orthodox and ultra-Orthodox Jewish rabbis emerge from Israel on a weekly basis. Religious leaders not usually known for proclamations of imminent prophetic fulfillment have been telling all who will listen to stay close to the Holy Land, avoiding travel abroad if possible, because Messiah's arrival could come at any moment. Messiah's appearance will culminate with Israel's restoration to its rightful place at the head of nations—and with the destruction of its enemies.

Secular influence on Western culture since the Enlightenment and the lasting impact of amillennial thinking in the Roman Catholic Church has drained expectations of unfulfilled prophecy from most mainline Christian denominations. Even here in the United States, home to more self-described Christians than any other nation on the planet (at least until 2030, when China takes the lead), most of us believers hear so little about Bible prophecy in church that what we do know is probably absorbed from pop culture—something about an Antichrist and a thing called Armageddon that destroys the world. Or was that the comet Bruce Willis blew up?

Hear this: The future of your eternal soul is far too important to draw guidance from Hollywood.

Prophecy teaching is so poor among American Christians that a group from the charismatic wing has gained some influence over conservative politicians while teaching a "victorious eschatology." We'll spend some time in this book examining Dominion Theology, a hash of unbiblical teachings that, broadly speaking, paints a picture of a church victorious that literally conquers the earth before Jesus will—or can—return.

These mutually exclusive end-time expectations did not arise by coincidence. If the apostle Paul is to be believed, this is a deliberate strategy by the principalities, powers, thrones, and dominions he warned us about. These evil intelligences are, like DiCaprio's character in *Inception*, manipulating reality by shaping our beliefs about the way the end of history will play out.

Mutually exclusive eschatological views of Christians, Jews, and Muslims will lead to bloody conflict as the true believers of the world's monotheistic religions follow what they believe is the will of God. In truth, many will be nothing more than pawns in the Enemy's game—a game that will result in the Antichrist's rise to power at the head of a global government.

And among the Antichrist's most ardent supporters will be Christians who have failed to check their reality against the one sure thing that can identify and reject the Enemy's dreamscape visions—the Bible.

INTRODUCTION

YOU'RE STANDING ON a battlefield. A war rages around you, although you can't perceive it—not directly, anyway.

The combatants are supernatural. On one side, the Creator of the universe: God. Yahweh, or YHWH, if you prefer. On the other, evil intelligences who chose rebellion over service long ago.

In former days, people were closer to the conflict than we are today. Or maybe they were just more sensitive to the struggle. Three thousand years ago, if the Bible is to be believed, people understood that nations were under the rule of gods. While people might believe in a pantheon, nations in the ancient Near East were championed by national deities—Chemosh for Moab, Molech for Ammon, Marduk for Babylon, Enlil for Sumer, Ashur for Assyria, Hadad for Aram, Ba`al for Canaan and Phoenicia, Osiris or Amun-Ra for Egypt, and so on.

Today, Christians have a cosmology that's been, well, dumbed down since the apostolic age. For the last 1,600 years or so, we've been taught that since God is the only true god, then the gods of the countries around ancient Israel were nothing more than imaginary beings represented by carved images.

1

That is, to be blunt, spectacularly incorrect.

The authors of the books of the Bible understood that there were many small-G gods. They knew that these supernatural entities had been assigned to the nations, but that Israel had been reserved for Yahweh. Their writings reflected this. It's in the Bible. We just haven't heard about it in church. For example, when God told Moses:

> For I will pass through the land of Egypt that night, and I will strike all the firstborn in the land of Egypt, both man and beast, and **on all the gods of Egypt I will execute judgments**: I am the LORD. (Exodus 12:12, ESV, emphasis added)

Or later, when we read:

> They set out from Rameses in the first month, on the fifteenth day of the first month. On the day after the Passover, the people of Israel went out triumphantly in the sight of all the Egyptians, while the Egyptians were burying all their firstborn, whom the LORD had struck down among them. **On their gods also the LORD executed judgments.** (Numbers 33:4, ESV, emphasis added)

Now, are we supposed to believe that God inspired Moses to write about His righteous judgments on lifeless bits of wood and stone?

The conflict between Yahweh and the gods who chose to rebel is most clearly seen in Psalm 82:

> God has taken his place in the divine council;
> in the midst of the gods he holds judgment:
> How long will you judge unjustly
> and show partiality to the wicked? *Selah*
> Give justice to the weak and the fatherless;
> maintain the right of the afflicted and the destitute.
> Rescue the weak and the needy;

deliver them from the hand of the wicked."
They have neither knowledge nor understanding,
they walk about in darkness;
all the foundations of the earth are shaken.
I said, "You are gods,
sons of the Most High, all of you;
nevertheless, like men you shall die,
and fall like any prince."
Arise, O God, judge the earth;
for you shall inherit all the nations!
(Psalm 82:1–8, ESV, emphasis added)

Traditional analysis explains away the use of the word "gods" in this psalm as references to human rulers. This makes less sense than a straightforward reading of the text.

Yahweh presides over a divine council that is tasked with carrying out His orders. Some of these supernatural beings disobeyed orders and were informed by God that their punishment would be to die "like men." This threat makes no sense if God was addressing a human audience.

The divine council concept is not new, although it may be to you. (It was to me when I first encountered it ten years ago.) Rather than build a case for it here, which would cover territory that's already been explored in greater depth by better minds, I recommend Dr. Michael S. Heiser's excellent book *The Unseen Realm: Recovering the Supernatural Worldview of the Bible*. In a nutshell, seventy *bene elohim* ("sons of God") were assigned to supervise the nations after God divided mankind at Babel. (You'll note that seventy ethnic groups are named in Genesis chapter 10, the Table of Nations.) They apparently decided to present themselves as gods to humankind, which required God's counter-move—reserving one nation for Himself that would one day produce the Savior.

Mountains have always been key in this drama. People have known since our earliest days that mountains are sacred, the abode of the gods. Why? Is it the mystery inherent in locations that are remote and inaccessible? Is it the awe inspired by their size and beauty? Is it simply that primitive humans noticed that mountains were tall and closer to heaven than where they lived?

No. Mountains are sacred because the original location of the divine council was on a mountain. The original home of humankind was also on that mountain. After rebellion broke out, the Fallen chose other mountains to call their own. And all of history is a chronicle of the long war that broke out on the original cosmic mountain.

The purpose of this book is to show that the accounts of this war aren't just contained in the Bible. Secular history records important information about this conflict, although most scholars don't see it for what it is.

The war between God and the gods has been fought in the supernatural realm and in our time-space domain by human agents of the warring parties. We'll look at some well-known events in the Bible in a way you probably haven't before—acts of war by God against gods that you've been told were imaginary.

We'll examine one of the people groups Joshua and the Israelites had to fight to claim their new homeland in Canaan. We'll document the connections between this group of people and the shadowy race of giants called the Rephaim, the occult center that was Babylon, the king of Egypt who was forced to let God's people go, the mythical race of gods called the Titans, and—possibly—a religion that has grown out of the greatest supernatural PSYOP in history.

You read that right: There are links between the Titans of Greek mythology, ancient Egypt, Babylon, and Canaan, and the prophesied Babylon the Great.

The nation in question showed up in the path of the Israelites just as God's chosen people began to establish themselves as a nation. The physical and spiritual descendants of this ancient enemy have been Israel's neighbors for 3,400 years, and they seek Israel's destruction even today.

We'll also examine the roles of two divine entities that have been particularly resilient in the long war, appearing again and again at key points in history. And we'll present what we think is a plausible scenario for the end times that's best described as a cosmic double-cross, a PSYOP that exploits our human biases and weaknesses—nationalism, racism, and a desire to take onto ourselves a role that God has reserved for the Messiah.

While this all sounds like what skeptics call "woo," it's not. This is based on documented history, theology, and linguistics. No fringe research here and only a little speculation, although the scholars whose work we've used would probably disagree with our conclusions. That's to be expected. We're applying a supernatural filter to offer an analysis of why history happened the way it did, and to demonstrate that the historical record supports the biblical account if you look at it the right way.

But mainly we want to emphasize that the supernatural war that's been fought since Eden is more obvious than you think. If you read the Bible with your spiritual eyes open, you'll see stories you've known since Sunday school a whole new way.

The command centers of this long war are mountains. The prophesied climactic battle at the end of history will be fought for control of one specific mountain. We will identify the key mountains of history and lay out a prophetic scenario that the Enemy may use to deceive the world into accepting their man as a savior.

As with the protagonists in the film *Inception*, the Fallen have planted false information in our minds to warp our perception of reality. Forget about end times prophecy; most American Christians don't believe *Satan* is real. The success of this PSYOP isn't surprising, considering how few churches in America teach end-times prophecy. Avoiding the topic plays right into the Enemy's hands.

To confuse things even further, the Fallen have introduced other religions into the end-times mix, most importantly Islam. Muslims will play a significant role in the unfolding drama. To be blunt, their prophecies are so twisted that when the final years of history arrive, Muslims will find themselves fighting the Antichrist—but for the wrong reasons. And their end won't be victorious global jihad, but as sacrificial victims to draw Jews and Christians into the greatest double-cross in history.

Meanwhile, Jews, who are still looking for a geopolitical savior, and a great many Christians, ignorant of Bible prophecy or taught an unbiblical strain thereof, will welcome the Antichrist with open arms.

Jesus didn't warn us of a great deception for nothing.

1

············

EDEN

EDEN WAS THE original holy mountain. Yes, Eden was a garden—a garden on a mountain. See Ezekiel 28:

> Son of man, say to the prince of Tyre, Thus says the Lord GOD:
>> "Because your heart is proud,
>> and you have said, 'I am a god,
>> **I sit in the seat of the gods,**
>> in the heart of the seas,'...
>> You were the signet of perfection,
>> full of wisdom and perfect in beauty.
>> **You were in Eden, the garden of God;**
>> every precious stone was your covering,
>> sardius, topaz, and diamond,
>> beryl, onyx, and jasper,
>> sapphire, emerald, and carbuncle;
>> and crafted in gold were your settings
>> and your engravings.

On the day that you were created
they were prepared.
You were an anointed guardian cherub.
I placed you; you were on the holy mountain of God;
in the midst of the stones of fire you walked.
You were blameless in your ways
from the day you were created,
till unrighteousness was found in you.
In the abundance of your trade
you were filled with violence in your midst, and you sinned;
so I cast you as a profane thing from the mountain of God,
and I destroyed you, O guardian cherub,
from the midst of the stones of fire."
(Ezekiel 28:2, 12–16, ESV, emphasis added)

Eden was where the divine council met, the "seat of the gods" on "the mountain of God." And the first humans were there, members of the divine council.

Although Ezekiel 28:1–19 is addressed to the king of Tyre, probably Ithobaal III, who reigned between 591 and 573 B.C., most commentators agree that these passages are aimed at the divine rebel in Eden. Some try to desupernaturalize the section and apply the verses to Adam, but verses 17 through 19 describe punishments that cannot refer to Adam's post-garden life ("I exposed you before kings, to feast their eyes on you"—*what* kings?).

The garden was essentially the earth's first temple, God's dwelling place on earth among humanity. We'll look at this concept in more depth in the chapter on Mount Sinai, but Ezekiel 28 makes the point in its condemnation of the rebel in the garden:

By the multitude of your iniquities,
in the unrighteousness of your trade
you profaned your sanctuaries.
(Ezekiel 28:18a, ESV, emphasis added)

The word translated "sanctuaries," *miqdash*, is the same word used to describe the tabernacle built by Moses per God's instructions. This verse, then, identifies Eden on God's holy mountain as sacred space, unique and set apart from all other places on earth—the place where Yahweh walked and talked with man. Human history from the moment Adam and Eve disobeyed God's command to refrain from eating the fruit of the Tree of the Knowledge of Good and Evil has been a chronicle of His plan of redemption—to bring humanity back to His mountain—and the Enemy's attempts to thwart it.

Eden was also where the Enemy first employed a PSYOP that's used repeatedly: "You will be as gods." Many false religions (and false teachings within the body of Christ) can be boiled down to that lie.

The Enemy keeps using it because *it keeps working*. The apostles contended with it in the first century, and this PSYOP is alive and well in the twenty-first. Doctrines from movements as diverse as speculative Freemasonry and the New Age movement are built around the idea that we're all divine if we'd just recognize the spark within.

Of course, that's a lie. Instead of godhood, Adam and Eve lost immortality, got kicked out of the garden, out of the divine council, and off the holy mountain. Modern seekers looking for the path to enlightenment will fare even worse.

So who was the serpent in the garden? Most of us assume it was Satan, but maybe not. The serpent isn't named in the book of Genesis. In fact, Satan wasn't even a personal name in the Old Testament.

Satan means "accuser," and it's written *ha-shaitan* in the OT. It's a title—*the* satan, so it actually means "the accuser." Think of it as a job title, like prosecuting attorney.

The adversary in the garden is the *nachash*, which is the word translated into English as "serpent." It's based on an adjective that means "bright or brazen, like shiny brass." The noun *nachash* can mean "snake," but it also means "one who practices divination."

In Hebrew, it's not uncommon for an adjective to be converted into a noun—the term is "substantivized." If that's the case here, *nachash* could mean "shining one." And that's consistent with other descriptions of the satan figure in the Old Testament.

For example, in Isaiah 14, the character called Lucifer in the King James translation, based on the Latin words chosen by Jerome (*lux* + *ferous*, meaning "light bringer"), is named in Hebrew *Helel ben Shachar*— "shining one, son of the dawn." (Interestingly, Šahar was a Canaanite deity, so a better translation of the verse is "Day Star, son of Dawn," which leads to some interesting speculation about the nature and origins of Helel.)

Now, consider this in Daniel 10:

I lifted up my eyes and looked, and behold, a man clothed in linen, with a belt of fine gold from Uphaz around his waist. His body was like beryl, **his face like the appearance of lightning, his eyes like flaming torches, his arms and legs like the gleam of burnished bronze**, and the sound of his words like the sound of a multitude. (Daniel 10:5–6, ESV, emphasis added)

Obviously, "shining one" is a pretty good description of the angel who had to battle the "prince of Persia," another supernatural being, to bring his message to Daniel.

About nine hundred years before Daniel, when the Israelites started complaining on their way out of Egypt (see Numbers 21:4-9), God sent *saraph nachash* ("fiery serpents"), to bite them.

Saraph is the root word of seraphim, which roughly means "burning ones." But the key point of these verses in Numbers 21 is that the Hebrew words *saraph* and *nachash* are used interchangeably, meaning that rather than "fiery serpents," the actual translation should read "*saraph* serpents."

Deuteronomy 8:15 praises Yahweh for bringing Israel through "the great and terrifying wilderness, with its fiery serpents," reinforcing the interchangeability of *saraph* and *nachash*.

Now, if the mental image of flaming snakes isn't weird enough, the prophet Isaiah twice referred to *flying* serpents (*saraph `uwph*, in Isaiah 14:29 and 30:6). And in his famous throne room vision, Isaiah saw:

…the Lord sitting upon a throne, high and lifted up; and the train of his robe filled the temple. **Above him stood the seraphim. Each had six wings: with two he covered his face, and with two he covered his feet, and with two he flew.** (Isaiah 6:1–2, ESV, emphasis added)

Again, the root word of seraphim is *saraph*, the same word translated "serpent" in Numbers and Deuteronomy. In fact, aside from the Isaiah 6 passage above, every single mention of seraphim in the Old Testament refers to serpentine beings!

The flying serpent was a well-known symbol in the ancient Near East, especially in Egypt. It would have been very familiar to the Israelites. The uraeus, a cobra standing on its coil with its hood extended, was a royal symbol of protection used by pharaohs and Nubian kings. Tutankhamun's death mask is an excellent example; the uraeus' hood is depicted with six distinct sections that look a lot like wings.

Of course, some scholars cite this as evidence that the Hebrews' understanding of seraphim was influenced by, or borrowed from, Egyptian cosmology. That's a common message from skeptics—Israel copied its religion from its neighbors. We'll deal with that later.

The bottom line is this: What Adam and Eve saw in the Garden wasn't a talking snake, but a *nachash*—a radiant, divine entity, very likely of serpentine appearance.

Now, since you're paying attention, you'll remember that the divine rebel in Eden, the *nachash* of Genesis 3, was called a guardian cherub in Ezekiel 28. As we just showed you, *nachash* and *saraph*, the singular form of seraphim, are interchangeable terms. But if the rebel in Eden was one of the seraphim, how could he also be one of the cherubim?

Good question. Cherubim are mentioned more frequently in the Old Testament than the seraphim. They are usually referenced in descriptions of the mercy seat on top of the Ark of the Testimony or carved decorations in the Temple built by Solomon. The exceptions are the cherubim who guard the entrance to Eden and the four cherubim

Ezekiel saw in his famous "wheel within a wheel" vision by the Chebar canal.

Most of us today have a mental image of cherubim that was shaped by artists in the Middle Ages—cute, chubby little boys with dinky wings who filled up the empty space in religious paintings. Nothing could be farther from the biblical and archaeological truth. Cherubim are seriously bad dudes you do not want to mess with. See Josh Peck's book *Cherubim Chariots*.

The cherubim of the mercy seat are usually shown as a matched pair of plainly recognizable angels perched on top of the ark with their outstretched wings touching in the middle. The Bible doesn't describe these cherubim, telling us only that they have wings and faces. Why? Apparently, everybody in the fifteenth century B.C. was familiar with what a cherub looked like, and they knew it was right and proper for them to serve as Yahweh's throne-bearers. You see, God appeared to men above the mercy seat, "enthroned on the cherubim." (See Numbers 7:89; 1 Samuel 4:4; 2 Samuel 6:2: Psalms 80:1 and 99:1; and Isaiah 37:16.)

But the cherubim that Ezekiel saw looked like something from a nightmare:

> This was their appearance: they had a human likeness, but each had four faces, and each of them had four wings. Their legs were straight, and the soles of their feet were like the sole of a calf's foot. And **they sparkled like burnished bronze.**
>
> Under their wings on their four sides they had human hands. And the four had their faces and their wings thus: their wings touched one another. Each one of them went straight forward, without turning as they went.
>
> As for the likeness of their faces, each had a human face. The four had the face of a lion on the right side, the four had the face of an ox on the left side, and the four had the face of an eagle.
>
> Such were their faces. And their wings were spread out above. Each creature had two wings, each of which touched the wing of

another, while two covered their bodies. And each went straight forward. Wherever the spirit would go, they went, without turning as they went.

As for the likeness of the living creatures, their appearance was like burning coals of fire, like the appearance of torches moving to and fro among the living creatures. And the fire was bright, and out of the fire went forth lightning.

And the living creatures darted to and fro, **like the appearance of a flash of lightning.** (Ezekiel 1:5–14, ESV, emphasis added)

While these living creatures aren't identified as cherubim in these verses, they are specifically called cherubim in Ezekiel 10.

So how do we read this? These creatures sound nothing like the shining serpentine seraphim. What's even more confusing is the description Ezekiel gives of another type of angelic being, the *ophanim*—the wheels that UFO hunters love to call spacecraft:

And I looked, and behold, there were four wheels beside the cherubim, one beside each cherub, and the appearance of the wheels was like sparkling beryl. And as for their appearance, the four had the same likeness, as if a wheel were within a wheel. When they went, they went in any of their four directions without turning as they went, but in whatever direction the front wheel faced, the others followed without turning as they went. And their whole body, their rims, and their spokes, their wings, and the wheels were full of eyes all around—the wheels that the four of them had. As for the wheels, they were called in my hearing "the whirling wheels."

And every one had four faces: the first face was the face of the cherub, and the second face was a human face, and the third the face of a lion, and the fourth the face of an eagle. (Ezekiel 10:9–14, ESV, emphasis added)

Wait—the *ophanim* had the face of a cherub *and* the face of a human? What's the difference? Why a cherub instead of an ox for the fourth face? Is there some connection between the cherub and the ox?

Well…maybe. The word "cherub" probably comes from the Akkadian *karibu* (the "ch" should be a hard "k" sound, although we English speakers don't usually say it that way). It means "intercessor" or one who "prays." The karibu were usually portrayed as winged bulls with human faces, and huge statues of the karibu were set up as divine guardians at the entrances of palaces and temples. This is like the role of the cherubim placed "at the east of the garden of Eden…to guard the way to the tree of life" (Genesis 3:24, ESV).

This is speculation, but it's possible that the divine rebel in Eden, the "anointed guardian cherub," once held the job of protecting the tree of life.

In fact, cherubim were the gold standard for guarding royalty in the ancient Near East. In Assyria they were called *lamassu*, and the Akkadians called them *shedu*. They were sometimes depicted as winged lions rather than bulls and they were often incorporated into the thrones of kings. So the function of the biblical cherubim—guarding the tree of life and carrying the throne of God—was entirely consistent with what the neighbors of the Israelites knew about these beings. Based on archaeological finds in the Levant (modern-day Syria, Lebanon, Jordan, and Israel), the scholarly consensus is that the cherub was probably more like a winged sphinx than a humanoid with wings.

Interestingly, the golden throne of Egypt's King Tutankhamun, who probably reigned within a hundred years of the Exodus, incorporated sphinxes *and* winged serpents—possibly the Egyptian equivalents of cherubim and seraphim. (Too bad for him they didn't protect the boy king past his eighteenth birthday.)

In other words, the presence of the cherubim in the Bible isn't an accident or an invention of the Hebrew prophets. The cherubim were known by different names by the other cultures of the ancient Near East, but they served a similar role in all of them. The cherubim were supernatural bodyguards for the throne of Yahweh, and their imagery was appropriated by earthly kings. A bit of hubris, no doubt encouraged as a PSYOP by the Enemy. Remember, "you shall be as gods."

One more note, although it goes way beyond what we're covering here: The four faces of the cherubim correspond to the four cardinal points of the zodiac—Man (spring equinox, Aquarius); Bull (summer solstice, Taurus); Lion (fall equinox, Leo); and Eagle (winter solstice, Eagle—the ancient form of Ophiuchus, which stands above Scorpio).

Again, this isn't surprising. Remember, Ezekiel was in Babylon, home of the greatest astrologers the world has ever known. (When a spectacular star appeared in the sky over Judea in 3 B.C., guess who deciphered its meaning? Babylonian astrologers.) Not only was Ezekiel seeing the divine guardians of Yahweh's throne, he saw astronomical symbolism that had far more meaning in that place and time than it does for us. Everyone who read about his vision in the sixth century B.C. would have known right off what he'd seen. Not a one would have assumed that the wheels or "living creatures" were UFOs or aliens.

So we've identified, as best we can, the nachash, one of the entities—gods, if you will—who was a member of the assembly on God's holy mountain. But what about the other gods? Who else was in Eden with God, Adam, Eve, and the nachash? What do we know about them?

Actually, more than you'd think. Unfortunately for us, English doesn't convey the full sense of the Hebrew words that describe the supernatural beings in the Bible. For example, our English word "angel" covers a range of entities—cherubim, seraphim, ophanim, malakim, bene elohim, and others in Hebrew, as well as archangels and Watchers. That's made it easier for scholars and theologians to get around the idea that multiple gods are clearly described in the Bible.

We know those gods were in the garden, or Yahweh would not have inspired Ezekiel to call Eden "the seat of the gods." And it's possible they're mentioned in Ezekiel 28, just not in the way we expect.

Scholars generally agree that Ezekiel 28 is linked to Isaiah 14, another account of the divine rebel being tossed out of Eden:

How you are fallen from heaven,
    O Day Star, son of Dawn!
How you are cut down to the ground,
you who laid the nations low!
You said in your heart,
I will ascend to heaven;
above the stars of God
I will set my throne on high;
I will sit on the mount of assembly
in the far reaches of the north.
(Isaiah 14:12–13, ESV)

If Ezekiel 28 and Isaiah 14 describe the same event, which is probable, we may have confirmation of other divine beings in Eden.

Remember that in the Ezekiel account, God tells the nachash how He destroyed the rebel "from the midst of the stones of fire." Compare that with what we discussed above about the brazen, glowing, or burning appearance of the beings encountered by Moses, Daniel, and Isaiah. And in Psalm 104:4, we read that God "makes his messengers winds, His ministers a flaming fire."

In the Isaiah 14 passage above, we also see a reference to the "stars of God." Scholars agree that "stars" in the Old Testament often refer to the *bene elohim* ("sons of God"). For example, when Yahweh rebuked Job for his lack of faith:

Where were you when I laid the foundation of the earth?
    Tell me, if you have understanding.
    Who determined its measurements—surely you know!
    Or who stretched the line upon it?
    On what were its bases sunk,
    or who laid its cornerstone,
    when the morning stars sang together

and all the sons of God shouted for joy?
(Job 38:4–7, ESV, emphasis added)

The divine rebel in Eden was cast out of the garden and the divine council for his pride and his desire to set his throne "above the stars of God"—sons of God who appear as beings of fire and light. If we read the passages in Ezekiel 28 and Isaiah 14 as consistent with one another, we can speculate without stretching the Scriptures too far: The stones of fire in Eden *were* the sons of God that the nachash wanted to rule from his own "mount of assembly in the far reaches of the north."

Apologists for the Bible often try to desupernaturalize the puzzling references to fiery, flying serpents by offering plausible explanations—plausible from a scientific, naturalistic, twenty-first-century perspective, that is. Some suggest that the fiery serpents of Numbers 21 were Israeli saw-scaled vipers, a dangerous venomous snake native to the Sinai Pninsula. Others claim that the verses are proof that dragons or pterodactyls were alive and present during the Exodus. Both suggestions miss the point by looking to the natural world to understand the supernatural.

The consequences of the rebellion in Eden were immediate and harsh:

The LORD God said to the serpent, "Because you have done this, cursed are you above all livestock and above all beasts of the field; on your belly you shall go, and dust you shall eat all the days of your life.

I will put enmity between you and the woman, and between your offspring and her offspring; he shall bruise your head, and you shall bruise his heel." …

Then the LORD God said, "Behold, the man has become like one of us in knowing good and evil. Now, lest he reach out his hand and take also of the tree of life and eat, and live forever—"

[T]herefore the LORD God sent him out from the garden of

Eden to work the ground from which he was taken. He drove out the man, and at the east of the garden of Eden he placed the cherubim and a flaming sword that turned every way to guard the way to the tree of life. (Genesis 3:14–15, 22-24, ESV)

Well-meaning Christians for generations have pointed to Genesis 3:14 as the moment in history when snakes lost their legs. To repeat: That misses the point by desupernaturalizing the story. God wasn't amputating snakes' legs; He was describing the punishment the nachash would suffer in figurative language. Even casual observers of the animal kingdom know that snakes don't eat dust.

What happened was this: The nachash was cast down from the peak of the supernatural realm, "full of wisdom and perfect in beauty," to be the lord of the dead. What a comedown! Isaiah 14 makes a lot more sense when you keep a supernatural worldview in mind:

> Sheol beneath is stirred up
> > to meet you when you come;
> > it rouses the shades to greet you,
> > all who were leaders of the earth;
> > it raises from their thrones
> > all who were kings of the nations.
> > All of them will answer
> > and say to you:
> > "You too have become as weak as we!
> > You have become like us!"
> > (Isaiah 14:9–10, ESV)

Keep these verses in mind because we'll come back to them later. "The shades" Isaiah mentioned are the Rephaim (root word *rapha*), a mysterious group mentioned several times in the Old Testament. The Rephaim weren't an invention of the Hebrews. They were well known to their neighbors. Again, we'll examine them more closely in the chapter on Mount Sinai.

For Adam and Eve, the banishment affected the two of them and all their descendants through the present day. Instead of living with God as members of His council, we humans have struggled for millennia to make sense of a world that often seems to make no sense. The memory of our brief time in the garden of God has echoed down the centuries, and it may be the source of our belief that mountains are somehow special, reserved for the gods.

The main takeaway of the chapter is this: Eden was a lush, well-watered garden "on the holy mountain of God," which was where Yahweh presided over His divine council. The council included the first humans. They walked and talked with the supernatural "sons of God" who, based on clues scattered throughout the Bible, were beautiful, radiant beings. At least some of them were serpentine in appearance.

The long war between Yahweh and the sons of God who rebelled is not just about control of the spirit realm, it's also about whether humanity will be restored to its rightful place "in the seat of the gods"—among the divine council on the Holy Mountain of God.

———

Life after Eden must have been a crushing disappointment for early humans, especially the First Couple. Forget about the burden of living under the curse—toiling to coax enough food from the ground to survive, the pain of bringing new life into the world, and all the rest. The realization that they had disappointed their Creator and condemned their children and their children's children until the end of time to lives apart from Yahweh must have been nearly unbearable.

The Bible gives us very little on the rest of their lives. We only know the names of three of their children: Cain, Abel, and Seth. There must have been others and at least two of them were girls, because Cain and

Seth both married and had children of their own. (See? The old question, "Where did Cain find his wife?" isn't that hard to answer.)

It's understood that secular archaeologists and historians won't agree with much of what we believe about human history. That's okay. We Bible-believing Christians don't reject science when we interpret data through a biblical lens. Science is the process by which we collect and record information to test theories about the way things are. Analysis is what we do with that information after it's collected. It's not the science we often question, it's the analysis.

Scholars do agree, however, that civilization emerged in the Fertile Crescent around 10,000 B.C. (Note: We're using dates that are generally accepted by a consensus of scholars so we don't get bogged down arguing about the timeline. That's outside the scope of what we're trying to do here.) Agriculture, cities, writing, trade, science, and organized religion all developed in a broad arc that stretched from Egypt through the Levant and down into Mesopotamia.

Curiously, the evidence uncovered thus far makes it appear that civilization in Sumer, today's southeastern Iraq, appeared fully formed with no preliminary steps. Some scholars speculate that earlier attempts at organizing a civilization might lie at the bottom of the Persian Gulf. The theory goes that Eden, from which all humans came, was in something called the Gulf Oasis, a lush valley watered by the Tigris and Euphrates Rivers, plus the Karun River from Iran and the Wadi Batin from Saudi Arabia. The Persian Gulf rose rapidly between 6000 and 5000 B.C., during the Neolithic Wet Phase, and as the Gulf moved northward, people moved ahead of it, leaving the evidence of their earlier settlements beneath the waves.

This civilization is called the Ubaid culture by scholars. That's not what the people who lived in it called it, of course; we don't know what they called themselves because they never invented writing. The Ubaid civilization got its name from Tell al-'Ubaid, a small settlement mound in southeast Iraq where famous archaeologists Henry Hall and Sir Leonard Woolley dug up the first bits of pottery from those people between 1919 and 1924.

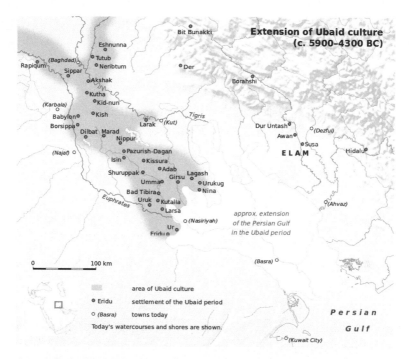

Map of Ubaid civilization sites in Sumer (credit John D. Croft / CC-BY-SA-3.0).

This happens a lot with prehistoric cultures. The names of their entire civilizations are accidents of discovery, coming from wherever the first archaeologist, bored shepherd, or greedy tomb raider found the first bit of evidence from people we'd never heard of before. Let's just hope that when archaeologists eight thousand years from now discover the remains of *our* civilization, it's not by stumbling across the shell of a Wawa convenience store or a Piggly Wiggly supermarket.

Archaeologists who study the Ubaid culture agree that it spread from Eridu in southeast Iraq, eventually going as far as what is today northwest Iran, northern Syria, southern Turkey, and the Levant (Syria/Lebanon/Jordan/Israel). The Ubaid civilization was typified by large, unwalled villages; rectangular, multiroom mud-brick houses; high-quality pottery; and the first public temples. Crop irrigation developed by about 5000 B.C., so cereals and grains could grow in the dry climate that again dominated the region. The first city in Mesopotamia, and therefore the oldest city in the world, appeared around 5400 B.C. Although agricultural settlements like Jericho (c. 9000 B.C.) and Jarmo, east of modern-day Kirkuk in Iraq (c. 7100 B.C.) are older, Eridu, located in what is today southeastern Iraq, was remembered by later Sumerians as the first city, with a degree of specialization among its citizens not seen before in other settlements.

The Sumerian King List, dated to about 2100 B.C., records it this way:

> After the kingship descended from heaven, the kingship was in Eridu. In Eridu, Alulim became king; he ruled for 28,800 years.[1]

Interestingly, the Bible may actually support this account.

> Cain went away from the presence of the LORD and settled in the land of Nod, east of Eden. Cain knew his wife, and she conceived and bore Enoch. **When he built a city, he called the name of the city after the name of his son, Enoch. To Enoch was born Irad,** and Irad fathered Mehujael, and Mehujael fathered Methushael, and Methushael fathered Lamech. (Genesis 4:16–18,ESV, emphasis added)

Some scholars, such as Egyptologist David Rohl, believe it's possible that the "he" in the second sentence refers to Enoch, not Cain. The last word, Enoch, might be a later addition. In that case, the builder of the city was Enoch, and the city was named for *his* son, Irad—hence, Eridu.

To speculate a little further, we can apply a rudimentary translation to the name Alulim and come up with "fourth man" (*A* = prefix + *lu* = "man" + *lim*, a contraction of *limmu* = "four"). Again, this is speculative and it may be way off base, so don't take it as gospel. But if it's correct, then Alulim might have been Irad, the "fourth man," or fourth generation, after creation—Adam, Cain, Enoch, Irad—and the first king of the first city on earth, Eridu—the city that bore his name.

Regardless of its origins, what is most interesting about Eridu is that besides being the oldest city in Mesopotamia and possibly in the world is that it was also the home of the oldest and largest ziggurat in Mesopotamia. This was the temple of one of the most important gods of the ancient Near East, one known as Enki to the Sumerians and Ea to the later Akkadians and Babylonians. Enki, the god of the sweet waters needed for life, was depicted with two streams of water flowing from his shoulders that represented the Tigris and Euphrates rivers, the main sources of fresh water in Mesopotamia.

Along with An (or Anu) the sky god and Enlil, the god of the air, Enki was one of the three most important gods in Sumer. He arrived very early in Sumer from Dilmun, probably the island of Bahrain in the Persian Gulf. In fact, the Sumerians believed Enki personally created Eridu, elevating it from the marshy ground on what was then the shore of the gulf.

Enki was the god of magic, craftsmanship, and wisdom. Although Enlil was the king of the gods, Enki was the keeper of the *mes* (sounds like "mezz"), decrees of the gods that formed the fundamental concepts and gifts of civilization—everything from religious practices to social interaction to music.

The Babylonian creation myth, the *Enuma Elish*, describes how everything on earth came into being through the defeat of the chaos goddess Tiamat by Marduk, son of Enki/Ea, the chief god of Babylon. However, the older Sumerian story credits Enki with giving life to all things, including mankind, and names Enlil the slayer of Tiamat.

The differences in the story are at least partly due to the ebb and flow of power over the centuries. Each city in Mesopotamia had a patron god or goddess. The importance of a deity was, as you'd guess, tied to the fortunes of its city. Just as Eridu was the home of Enki, Enlil was chief deity at Nippur, Inanna (Ishtar) was supreme at Uruk, the sun god Utu was the patron deity of Sippar, and so on. To give you an idea of the incredible amount of time we're dealing with, Enki had ruled in Eridu for about 3,500 years (!) before Marduk replaced Enlil at the head of the Mesopotamian pantheon, an event linked to Babylon's emergence as the region's dominant power in the eighteenth century B.C.

For context, that's about the same amount of time that's passed between Moses leading the Israelites out of Egypt and you reading this sentence.

This chapter is not in any way a thorough review of life, culture, or religion in ancient Mesopotamia, but there is one more aspect of life in the ancient Near East to call to your attention. It's something we only hear about from fringe pseudo-scholars who blame the phenomenon on extraterrestrials. Scholars—archaeologists and sociologists—have known at least since the late 1940s that people throughout Mesopotamia, before they learned how to write, figured out how to turn their children into coneheads.

It appears, based on human remains dated to between about 10,000 B.C. and 3500 B.C., that cranial deformation was widespread in the Ubaid culture, and Eridu—the world's first city, possibly built by Cain or his son—was Ground Zero for head shaping. An archaeological dig at Eridu just after World War II discovered about a thousand bodies that were buried during the Ubaid. Of the 206 sets of remains the archaeologists exhumed, "all of the crania had been deformed in one fashion or another."[2]

Got that? 206 out of 206. Not a few, and not just the elites. It appears that everybody from every strata of the Eridu culture had a deformed skull.

Now, instead of asking why, the lead archaeologist decided "earth pressure" after burial was the cause—even though none of the skulls were cracked or broken, which would be expected if the deformations had occurred after death.

Evidence of head shaping has been found at sites all over Iraq, southwestern Iran, eastern Turkey, the valleys of the Zagros mountains, and the western shores of the Persian Gulf, dated from 7500 B.C. to about 4000 B.C. After that, the practice seems to disappear.

Hmm. If we place a global flood sometime between 4000 and 3500 B.C....

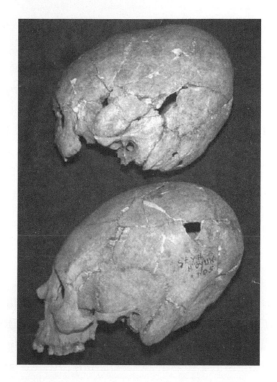

Deformed crania from the Ubaid site at
Seyh Höyük in Turkey (photo by Kirsi O.
Lorentz).

The big question is why this was even a thing. Please understand that
we're not suggesting that these were genetic mutations or part-human
Nephilim, the angel-human hybrids mentioned in Genesis 6 (although
the Nephilim would have been around during at least part of this time).
But who wakes up one morning in 7000 B.C. and decides to wrap
something around the baby's skull to see if it makes his head pointy?

What inspires that? And why was Eridu the starting point for this?

A study published in the academic journal *Paléorient* in 1992 concluded
that the practice of head-shaping, which is found around the world, must
have originated in the Near East because it was so widespread there.
However, the researchers believed the deformation was not necessarily
intentional, but probably "incidental to patterns of head-gear."[3]

Really? For more than six thousand years, our ancestors *accidentally* forced their babies to wear head gear so tight it *deformed their skulls*?

Here's another bit of data to chew on. At Eridu and nearby sites in ancient, pre-Flood, southern Sumer, and *only* there, archaeologists have found about 120 terracotta figurines scholars call ophidian. That's fancy talk for snake-like. They're slender bipeds, adorned with button-like protuberances, more often female than male, and often in poses that are exclusively mammalian—for example, a female lizard-like figure suckling an infant.

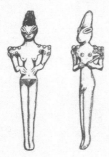

Ophidian figurines made from terracotta and bitumen. Female found at Ur in 1931, male at Eridu in 1948–49. Drawing of female based on object #31-16-733 at Penn Museum; male figurine was on display at the National Museum of Iraq.

Now, the ancient astronaut evangelists on cable TV shows have an answer for this. They were our space ancestors, they say, the Anunnaki who came from the stars to create humanity from ape DNA.

They're thinking along the right lines, but because they ignore the supernatural, they miss the more likely answer.

The aforementioned scholars pointed out in their paper that there had been no serious study of those figurines and what they meant to the ancients, and no scholarly literature on the origins of human cranial deformation (at least as of their writing in 1992).

Why might that be?

While there hasn't been much scholarly attention paid to the snaky figurines, there have been several papers published within the last ten years on head-shaping in the ancient Near East. Still no conclusions on why or how it got started, but it's clear that the people who lived in the region—descendants of the refugees from Eden—made a habit of this odd practice.

We'll never know for sure, but we can speculate: The people who formed the earliest human civilizations copied a look that someone, somewhere, had seen and decided was a physical ideal. What motivates modern-day teens to wear the clothes and hairstyles they do? Except that this *particular* fashion statement didn't change with the seasons—it appears to have been worn by nearly everybody for more than six thousand years!

It's not likely that this was a simple fashion statement. However it started, this was apparently a practice that was believed to convey some advantage.

Think back to what we discussed about the serpentine nachash and seraphim, and remember that at least one of them rebelled against Yahweh. Is it possible that the citizens of the prehistoric Near East were trying to curry favor with a god?

———————

Now, if we haven't already, here is where we may run off the rails. But let's run through one more theory, look at the evidence, and see if it makes sense.

As mentioned earlier, Eridu is considered by archaeologists to be the center of the Ubaid culture. The Ubaid period is defined as the civilization in the ancient Near East just before the time of Nimrod and the Tower of Babel, roughly between 6500 B.C. and 3800 B.C. It would have begun not long after Adam and Eve got kicked out of the garden.

Now, God wanted humanity to "have dominion over the fish of the sea and over the birds of the heavens and over every living thing that moves on the earth" (Genesis 1:28). It was not for His people to take dominion over each other. As Jesus told His disciples thousands of years later, "You know that the rulers of the Gentiles lord it over them, and their great ones exercise authority over them. It shall not be so among you" (Matthew 20:25–26a).

Just as Adam and Eve were created to work the land, God's design for humans was to be more or less self-sufficient—growing our own food, tending our own flocks, and helping each other whenever and however we're needed. When you live that kind of life, you're too busy to lord it over your neighbors. And, frankly, since you're not depending on hand-outs for your family's daily bread, it's harder for government to lord it over you. It makes sense that Israel was led by judges in times of trouble during its early years, but a king wasn't part of the original plan—even though God, who knows the end from the beginning, surely saw what was coming and let the Hebrews make Saul king anyway.

Earlier, we described the main features of the Ubaid civilization. Archaeologists and sociologists have noticed two other things that emerged during the Ubaid period—first, it was a transition from a rural to an urban society, and second, society became increasingly stratified.

In other words, the evidence—things people were buried with, mainly—shows that as people moved from the country to the city, the rich got richer and the poor got poorer.

This period also saw the construction of the first temples in Mesopotamia. Each city apparently had a temple for its own local god. And each temple had a granary for collecting the offerings of the commoners. Of course, this means somebody oversaw who gave how much, and—more importantly—who *received* how much. Archaeologists and sociologists believe this led to an elite class of hereditary leaders, who may have been the ones doling out the grain from temple granaries.

Substitute income tax and social security control over the grain harvest and rationing, and it's clear that things haven't changed all that much in the last eight thousand years. But I digress.

It's fair to ask how this situation came about. We can only speculate since we don't have any written records to study. Logically, though, it's a pretty good guess that entities who'd rebelled against Yahweh at this early stage had set themselves up as gods. It's conceivable, based on what we know about angels from the Bible, that one or more of them appeared to the pre-Sumerians and encouraged certain candidates who were receptive to the idea that 1) the gods needed temples, and 2) the temples required priest-kings to oversee offerings and ensure that the gods were pleased. The result was a civilization in which freedom was restricted to the upper class while the peasants worked the land to support them.

We're speculating, of course, but from the standpoint of the Fallen, breaking civilization into classes was a brilliant PSYOP. Destroying the world is much easier when you only have to manipulate a handful of useful idiots who can convince the rest to do things your way.

Scholars who study the Ubaid period have concluded that it was a time of increasing disparity between the classes, but they don't look at the situation through a spiritual lens. It happened, but why? They analyze factors like climate and technology, but they ignore the possibility that the unseen realm may have guided those changes.

God's original design was for people to work the land and live free. As Christians, who should accept the existence of the spirit realm by default, it should be natural for us to ask whether the lesser elohim might have rebelled against Yahweh's plan for creation, and if so, how that would play out. If we don't, you can bet secular historians, archaeologists, and anthropologists won't bother.

Since the trend during the Ubaid period was toward ever-larger villages with temples of the gods at the center of everyday life, contrary to God's command to "multiply and fill the earth," it's a fair guess that this change was the work of entities who were disloyal to their Creator—and who wanted to destroy the work that He'd called very good.

SUMMING UP

Humanity was created to play a role in God's divine council, which met on His holy mountain, Eden. That plan was disrupted when Adam and Eve, deceived by the nachash, disobeyed God's command and ate the fruit of the tree of the knowledge of good and evil. As punishment, humanity was kicked off the mountain and out of the council while the nachash was demoted from the pinnacle of beauty and wisdom to serving as lord of the dead.

Archaeology shows that human civilization appears to have appeared in ancient Sumer, fully formed. As the Ubaid culture spread from the first city, Eridu, in what is today southeastern Iraq, a class society developed with the elites linked to the temples that existed in every city in Mesopotamia. Oddly, it appears that head-shaping was standard practice in this society for thousands of years, a practice that was reflected even in the small ophidian figurines discovered at Eridu. While cranial deformation has been practiced by other cultures in more recent years, it's often used to distinguish one class from another. But it looks like the Ubaidians were *all* doing it, based on the remains that have been found.

And those figurines raise more questions. What were those serpentine humanoid figurines supposed to look like, and why were they so popular at Eridu and the nearby city of Ur? Is it possible that the prehistoric people of the ancient Near East were trying to emulate the appearance of the nachash?

2

HERMON

WHILE THE EARLY Sumerians were figuring out how to grow crops without rain (and turn their children into coneheads), interesting things—historic things—were happening far to the northwest. The second peak in our chronological list of spiritually strategic holy mountains is Mount Hermon.

Hermon is the highest, most majestic peak in the Levant. At 9,200 feet above sea level, it dominates the Golan Heights on the border between Israel and Syria, anchoring the southern end of the Anti-Lebanon Mountains. It has been considered sacred for most of human history.

Mount Hermon was a holy site as far back as the old Babylonian period, nearly two millennia before Christ, and maybe even earlier. In the Old Babylonian version of the Gilgamesh epic, which dates to the eighteenth century B.C. (roughly the time of Jacob), "Hermon and Lebanon" were called "the secret dwelling of the Anunnaki." The Ninevite version of the poem, written about six hundred years later, describes the monster slain by Gilgamesh, Humbaba (or Huwawa), as the guardian of "the abode of the gods."[4]

The Anunnaki were the seven chief gods of the Sumerian pantheon:

Anu, the sky god; Enlil, god of the air; Enki, god of the earth; Ninhursag, mother goddess of the mountains; Inanna (Babylonian Ishtar), goddess of sex and war; Sîn, the moon god; and Utu, the sun god. They are mentioned in texts found in what is today southeastern Iraq that date back to the twenty-seventh century B.C. So it's possible the more recent versions of the Gilgamesh story from Babylon and Nineveh remember more ancient traditions. And we'll discuss later why those ancient traditions may have been brought to Babylon from Syria, far to the west.

The name "Hermon" appears to be based on a root word that means "taboo," similar to the Hebrew word *kherem*, or "devoted to destruction." The word is often translated into English as "under the ban."

The first appearance of the word in the Bible is Exodus 22:20: "Whoever sacrifices to any god, other than the LORD alone, shall be **devoted to destruction** (*kherem*)." But this condemnation, or "the ban," wasn't just invoked against disobedient Israelites. Some of the inhabitants of Canaan were also declared *kherem* by Yahweh—specifically those who were known to be giants, or at least descended from giants.

That begs the question: Where did the giants come from? A curious episode is recorded in the first four verses of Genesis chapter 6:

> When man began to multiply on the face of the land and daugh-ters were born to them, the sons of God saw that the daughters of man were attractive. And they took as their wives any they chose. Then the LORD said, "My Spirit shall not abide in man forever, for he is flesh: his days shall be 120 years."
>
> The Nephilim were on the earth in those days, and also after-ward, when the sons of God came in to the daughters of man and they bore children to them. These were the mighty men who were of old, the men of renown. (Genesis 6:1–4, ESV)

Scholars have debated the meaning of the term "Nephilim" for millennia. Most believe it comes from a Hebrew root, *napal*, meaning "to fall" or "cast down"—literally, "fallen ones."

However, Bible and ancient language scholar Dr. Michael S. Heiser (author of the excellent book *The Unseen Realm*, which is highly recommended), at his website www.SitchinIsWrong.com (a reference to the late Zecharia Sitchin, who popularized the idea that the Sumerian gods, the Anunnaki, were extraterrestrial astronauts who created humanity from ape DNA), contends that this cannot be the case:

> The form *nephilim* cannot mean "fallen ones" (the spelling would then be *nephulim*). Likewise *nephilim* does not mean "those who fall" or "those who fall away" (that would be *nophelim*). The only way in Hebrew to get *nephilim* from *naphal* by the rules of Hebrew morphology (word formation) would be to presume a noun spelled *naphil* and then pluralize it. I say "presume" since this noun does not exist in biblical Hebrew—unless one counts Genesis 6:4 and Numbers 13:33, the two occurrences of *nephilim*—but that would then be assuming what one is trying to prove! However, in Aramaic the noun naphil(a) does exist. It means "giant," making it easy to see why the Septuagint (the ancient Greek translation of the Hebrew Bible) translated *nephilim* as *gigantes* ("giant").[5]

In short, the Jewish scholars who translated the Old Testament into Greek about two hundred years before the birth of Jesus clearly understood that the Nephilim were giants, not just men who "fell away" from God.

Likewise, the Hebrew words translated "sons of God" in the passage, *bene elohim*, refer to divine beings, not mortal men. Now, that hasn't been the consensus among Christian scholars since about the fifth century, thanks to the great theologian Augustine. He popularized the "sons of Seth" theory to explain away the weird supernatural element of the passages above. In short, the Sethite view is that the sons of God were men from the godly, righteous line of Seth who began intermarrying with women from the corrupt, wicked line of Cain.

Frankly, this defies logic on several points:

1. How likely is it that all the Sethite men were good while all the Cainite women were bad?
2. We're supposed to believe Cainite men never married Sethite women?
3. Why would these unions produce Nephilim, understood to be giants by Jewish rabbis and early Christians alike?
4. Why would these unions lead to wickedness so great that God had to wipe out everything that walked the earth except Noah, his family, and the creatures in the ark?
5. Every other use of *bene elohim* in the Hebrew Scriptures refers to divine beings.

Problems with the supernatural understanding of the text usually focus on whether angels and humans could successfully produce children. Proponents of the Sethite view often point to Jesus' teaching on the resurrection of the dead:

> For in the resurrection [people] neither marry nor are given in marriage, but are like angels in heaven. (Matthew 22:30, ESV)

The key words are "in the resurrection" and "in heaven." Noah's neighbors were flesh and blood, not resurrected, and the angels who "came in to the daughters of man" were most definitely not in heaven.

There are several examples in the Bible of divine beings interacting with humans in physical ways—eating, drinking, and even engaging in a dustup in front of the house of Lot (Genesis 19:5-11). Why couldn't they procreate as well?

The final nails in the coffin of the Sethite view are the references to this event in the New Testament. Both Peter and Jude refer to the only example in Scripture where angels transgressed:

> For if **God did not spare angels when they sinned**, but cast them into hell and committed them to chains of gloomy darkness to be

kept until the judgment; if he did not spare the ancient world, but preserved Noah, a herald of righteousness, with seven others, when he brought a flood upon the world of the ungodly; **if by turning the cities of Sodom and Gomorrah to ashes he condemned them to extinction**, making them an example of what is going to happen to the ungodly. (2 Peter 2:4, ESV, emphasis added)

And the **angels who did not stay within their own position of authority, but left their proper dwelling,** he has kept in eternal chains under gloomy darkness until the judgment of the great day—**just as Sodom and Gomorrah and the surrounding cities, which likewise indulged in sexual immorality and pursued unnatural desire,** serve as an example by undergoing a punishment of eternal fire. (Jude 6–7, ESV, emphasis added)

If there was any doubt about what the angels did that deserved punishment, Peter and Jude clarified things by specifically identifying the sin of the angels as sexual by linking it to the sins of Sodom and Gomorrah. Crossing the species barrier between angel and human is just as taboo as the barrier between human and animal.

It is significant that the phrase translated "cast them into hell" in 2 Peter 2:4 is the Greek word *tartaroo*, a verb meaning "thrust down to Tartarus." This is the only time in the New Testament that the word is used, meaning it requires special attention. Tartarus was separate from Hades, a place of torture and torment even lower than Hades in Greek cosmology. It was believed to be as far below Hades as the earth is below Heaven. And Peter, under the inspiration of the Holy Spirit, chose that word to describe the punishment reserved for the angels who had engaged in illicit sexual relations with human women. Stick a mental bookmark there because we'll have to come back to this.

The extrabiblical books of Enoch and Jubilees expand on the story, adding detail and context that's not in the Bible. Mount Hermon is where two hundred Watchers, a class of angelic beings mentioned in chapter 4 of the book of Daniel, descended and began cavorting with human women. From these unions came the Nephilim, the giants of Genesis 6.

The Watchers, according to Enoch, were led by Semjâzâ, who was apparently worried that he'd take the fall for what they were about to do:

> And Semjâzâ, who was their leader, said unto them: "I fear ye will not indeed agree to do this deed, and I alone shall have to pay the penalty of a great sin." And they all answered him and said: "Let us all swear an oath, and all bind ourselves by mutual imprecations not to abandon this plan but to do this thing." Then sware they all together and bound themselves by mutual imprecations upon it. And they were in all two hundred; who descended [in the days] of Jared on the summit of Mount Hermon. (The Book of Enoch 6:3–6a, R. H. Charles translation)

The trade offered by the Watchers was knowledge, just as it was in Eden. In exchange for the pleasures of the flesh, Semjâzâ and his minions offered charms and enchantments, astrology, the art of making weapons, cosmetics, and writing, among other things—presumably arts humans would have developed or discovered over time rather than downloaded on society all at once.

However, the giant offspring of these unholy unions, the Nephilim, pillaged the earth and endangered humanity. They consumed everything men had. When that wasn't enough, they began eating people and even each other. Enoch describes the giants as creatures of insatiable desire who threatened to terminate the bloodline of the future Messiah by violence—and, apparently, by corrupting the human genome.

The Book of Jasher suggests that the transgression of the Watchers and the Nephilim went beyond corrupting humankind, including "the mixture of animals of one species with the other" (Jasher 4:18).

We can only speculate what that was about. Was this where the legends of chimeric beings like centaurs and satyrs began? Or were they legends at all? But we do know that Yahweh sent a flood that rid the earth of all flesh except for the eight people and the animals aboard the ark. The Watchers who started it all, according to Peter and Jude, are chained in Tartarus, and they'll stay there until "the judgment of the great day."

But we're still dealing with the consequences of the sin of the Watchers. You see, the giants are still with us today. The Nephilim were the origin of demons.

> **And now, the giants, who are produced from the spirits and flesh, shall be called evil spirits upon the earth, and on the earth shall be their dwelling.** Evil spirits have proceeded from their bodies; because they are born from men and from the holy Watchers is their beginning and primal origin; **they shall be evil spirits on earth, and evil spirits shall they be called.** [As for the spirits of heaven, in heaven shall be their dwelling, but as for the spirits of the earth which were born upon the earth, on the earth shall be their dwelling.] And the spirits of the giants afflict, oppress, destroy, attack, do battle, and work destruction on the earth, and cause trouble: they take no food, but nevertheless hunger and thirst, and cause offences. **And these spirits shall rise up against the children of men and against the women, because they have proceeded from them.** From the days of the slaughter and destruction and death of the giants, from the souls of whose flesh the spirits, having gone forth, shall destroy without incurring judgement. (1 Enoch 15:8–12, 16:1, R. H. Charles translation, emphasis added)

This may be new to you, but it's not a new concept. The church today doesn't deal much with the topic of demons, but it's clear that the early church understood that they were real and distinct from angels.

Messengers from God in Scripture are described as men. Other classes of angel, like the nachash, cherubim, and seraphim we discussed earlier, are obviously entities of a whole different kind, but always possessing physicality—an outward appearance that can be observed and described by the prophets.

Demons, on the other hand, require a host. They are spirits only, at least in biblical accounts. It was the consensus view among Jews and Christians until the time of Augustine in the late third/early fourth century that demons were what remained of the Nephilim after the Flood.

"And when the angels of God saw the daughters of men that they were beautiful, they took unto themselves wives of all of them whom they Chose." Those beings, whom other philosophers call demons, Moses usually calls angels; and they are souls hovering in the air. (*Philo,* On the Giants 6)

In my opinion, however, **it is certain wicked demons, and, so to speak, of the race of Titans or Giants, who have been guilty of impiety towards the true God**, and towards the angels in heaven, and who have fallen from it, and who haunt the denser parts of bodies, and frequent unclean places upon earth, and who, possessing some power of distinguishing future events, because they are without bodies of earthly material, engage in an employment of this kind, and desiring to lead the human race away from the true God. (*Origen,* Against Celsus 4.92, emphasis added)

God…committed the care of men and of all things under heaven to angels whom He appointed over them. But the angels transgressed this appointment, and were captivated by love of women, and begot children who are those that are called demons; and besides, they afterwards subdued the human race to themselves, partly by magical writings, and partly by fears and the punishments they occasioned, and partly by teaching them to offer sacrifices, and incense, and libations, of which things they stood in need after they were enslaved by lustful passions; and among men they sowed murders, wars, adulteries, intemperate deeds, and all wickedness. (*Justin Martyr,* 2 Apology 5)

Justin Martyr not only understood that the Nephilim were the source of the demons that plague mankind, he also clearly knew that the rebellious members of the divine council were the false gods of the pagan world.

So it's obvious that Mount Hermon's role in the history of the world didn't end with the Flood of Noah. Not only did the death of the Nephilim result in a plague of demons, which Jesus confronted during His ministry, another group of fallen bene elohim set up shop on the mountain after the showdown at Babel.

According to Belgian scholar Edward Lipinski, Hermon was known to the ancient world not only as the secret dwelling place of the Anunnaki, but also as specifically the mountain of the divine assembly of the northwest Semitic god El, the creator god of their pantheon. [6]Mount Hermon is where El held court with his consort Asherah and the "seventy sons of El."

Remember that number, seventy. We will see it again.

El was a name that came to be used in Hebrew as a generic term for "god"—El, Elohim, El Elyon, etc. It's possible that the epithet El Shaddai, possibly meaning "god of the mountain," was applied first to El. Another Mesopotamian god, Amurru, was called "Bel Šade," or "lord of the mountain." (The š sounds like "sh.") This was another PSYOP—appropriating a name by which Yahweh identified Himself to confuse things. In Exodus, Yahweh told Moses that it was by the name El Shaddai, usually rendered "God Almighty" in our English Bibles, that He introduced Himself to Abraham, Isaac, and Jacob (Exodus 6:3). That's why Jacob became Isra-el, not Isra-yahu.

Of course, skeptics take this to mean that Jews and Christians are confused about who we worship. It's actually El, they claim, and the followers of Yahweh are so dense we've gotten it wrong for the last 3,500 years. In point of fact, however, it's the skeptics who have fallen for another PSYOP by the Enemy; they don't care what we believe as long as it's not the one thing that's true.

But make no mistake: El of the Canaanites was not Yahweh of Israel, and Yahweh was not El. In the Canaanite pantheon, El was a figurehead.

Real power was wielded by Ba`al, the king of the gods. In the Canaanite myths, there was a power struggle between the gods over who would hold that title, and El didn't seem to have the power or the will to just pick a winner. Neither did he seem all that interested in ruling himself. If you had to pick a word to describe El, it would be "semi-retired."

That is definitely not the God of the Bible.

But back to Mount Hermon. The PSYOP of the Watchers, not surprisingly, was the same one that worked in Eden: "You will be as gods." Apparently, the promise made by these "angels who…left their proper dwelling," according to Jude, was free access to women in exchange for information on making weapons and cosmetics, the secrets of sorcery and concocting potions—even reading, writing, and 'rithmetic. But it was for the Nephilim that Yahweh intervened.

Interestingly, long before the Hebrews were recognized as a distinct people, the Mesopotamians knew about the Watchers. They called them *apkallu*. They were created by the god Enki to bring knowledge and the gifts of civilization to mankind. If you're thinking right about now that this sounds a lot like the Watchers (and a bit like the Greek myth of the Titan Prometheus), you're on the right track.

According to the myths, there were seven apkallu before the Flood and four afterward. The word "apkallu" comes from the Sumerian *ab* (water), *gal* (big), and *lu* (man). They were considered only *partly* evil—occasionally dangerous and capable of malicious witchcraft. They were usually depicted as human in form but with wings, or sometimes as hybrid bird-man or bizarre fish-man creatures.

The antediluvian apkallu were divine like the Watchers. In one story from the Babylonian period, the *Epic of Erra*, the god Marduk banished the apkallu to the *abzu*, the fresh water aquifer below the temple of Enki, and told them never to return as punishment for provoking him to send the flood.

Hmm. Supernatural beings linked to a global flood, afterward banished to the abyss. Sound familiar?

Cylinder seal image showing two Mesopotamian sages called **apkallu**, the human-headed winged type, and the fish-cloaked type. Scholars identify the apkallu with the biblical Watchers.

Interestingly, the four apkallu who appeared after the flood were only partly divine. Being "of human descent," they could mate with humans—like the Watchers.

The last of the apkallu, Lu-Nanna, was "two-thirds apkallu." This matches the status of Gilgamesh, who was described two-thirds divine and one-third human. On one cylinder seal, Gilgamesh is called "lord of the apkallu," and elsewhere he's credited with bringing back knowledge that existed before the Flood.

Scholars who have made the connection between the apkallu and the Watchers tend to see the way the Watchers are portrayed in Jewish literature from the Second Temple period, like the Book of Enoch, as a Jewish response to the Babylonian captivity. The apkallu, although potentially dangerous, had preserved secret pre-Flood knowledge, which was prized by the pagan wizards of Babylon. To the Jews, however, such knowledge was evil, and the Watchers were portrayed accordingly in Enochian texts.

But those scholars evaluate the situation through secular lenses, analyzing the texts apart from any spiritual reality. Looking back through more than five thousand years, we can only make educated guesses about most of what happened, but we can assume that Noah and his family knew about the Watchers and the horrors they'd unleashed on the world. It's no surprise that memories of the Watchers survived the Flood, although the stories obviously changed over time. That was probably due in part to the confusion of languages after Babel, but also to justify a return to the occult practices the Watchers taught humanity before the Flood.

One of the fascinating aspects of the story of Gilgamesh is that archaeologists generally consider him a real, historic character. In 2003, a team digging at the site of ancient Uruk believed they'd found the tomb of Gilgamesh beneath what was the former course of the Euphrates River— but their discovery came a month after the United States military invaded Iraq in 2003, which put a stop to the dig.

Scholars have known for years that there are parallels in Mesopotamian legend and the biblical accounts of the patriarchs. Enoch is similar to an antediluvian king named Enmeduranki, and Noah is variously called Utnapishtim (Babylon), Ziusudra (Sumer), and Atra-Hasis (Akkad), depending on which culture wrote the story. But even those accounts are part of a supernatural PSYOP. For example: The accounts from Mesopotamia portray Gilgamesh as a mighty warrior, a hero, two-thirds god and one-third man. He has adventures and slays monsters, notably Humbaba, or Huwawa, the defender of the faraway cedar forest who'd been assigned to terrorize humans by the god Enlil.

In the Second Temple Jewish account we call the Book of Giants, Gilgamesh was himself one of the gigantic offspring of the Watchers, as was Humbaba, the monster Gilgamesh set out to kill. This is how Gilgamesh was viewed by Jews between the time of the Babylonian captivity and the birth of Jesus—basically, he was one of the Nephilim.

As noted above, Humbaba might also have been pronounced "huwawa." Dr. David Livingston, the founder of Associates for Biblical Research, points out that Huwawa sounds a lot like Yahweh. If he's right, then it's possible we've discovered another Enemy PSYOP: The *real* mission of Gilgamesh, as the Fallen wanted Mesopotamians to understand it, was to kill the monstrous guardian of the secret home of the gods—Yahweh.

Adam's story is a little different. In the Mesopotamian accounts, Adapa is one of the apkallu, the son of the great god Enki. One day, in a rage, Adapa broke the wings of the south wind for overturning his fishing boat. He was called to the home of the gods by the sky god, Anu, to explain himself. Before his journey, Enki warned Adapa not to accept any food or drink because Any meant him harm. Adapa followed instructions, accepting only a robe and anointing oil, and missed his chance at immortality by refusing to eat the food of the gods.

In the Bible, of course, the story is reversed: Adam was created as an immortal by Yahweh. Adam and Eve *lost* their immortality by disobeying the command *not* to eat from the Tree of the Knowledge of Good and Evil.

The account of the flood is quite different, too. The Babylonian hero, Utnapishtim—who reigned from the city of Shuruppak as the last king of Sumer before the flood—was secretly warned about the coming deluge by Enki. The chief god, Enlil, was angry because humans made so much noise that he couldn't sleep, and so he decided to solve the problem by killing them all. Enki, displaying the guile that made him a match for the more powerful Enlil in Sumerian myths, shared the high god's plan with a wall that Utnapishtim just happened to be standing beside, so that Enki could warn the human without disobeying Enlil's command to keep the coming flood a secret.

Of course, the biblical account is much different. Yahweh wiped the earth clean because of humanity's evil, corrupted in body and spirit by the Watchers. Noah, who was "perfect in his generations," was spared with his family by Yahweh to repopulate the earth.

There are two levels to this PSYOP: First, the villain of these stories is the chief god of the pantheon. Enlil, "lord of the air." wanted to keep humans stupid, subservient, and quiet. But since the annoying humans wouldn't cooperate, he'd commit genocide so he could get some sleep. The hero of the story is Enki, who gave Sumerians the secrets of civilization—a sort of proto-Prometheus. He preserved humanity through the flood by cleverly obeying the letter but not the intent of Enlil's order not to warn humans that it was coming.

So generations of people in the Fertile Crescent grew up hearing about how the clever Enki saved humanity from the angry, selfish chief god Enlil, who, not coincidentally, bears a passing resemblance to Yahweh.

The second level of the PSYOP has only revealed itself within the last hundred and fifty years or so as these stories were rediscovered by archaeologists. Scholars and skeptics have concluded that since the Babylonian tablets are older than the book of Genesis, the Hebrews must have copied their religion from Mesopotamia.

Of course, they ignore the differences. For example, the gods of Sumer and the rest of the ancient world were not all-powerful. Each one had a specific area of influence. They were all created beings (except for the earliest primal gods like Uranus and Gaia, who were obviously themselves not all-powerful), subject to the same follies and petty jealousies as humans, and they could die.

Yahweh, on the other hand, exists outside of time. He existed before anything and spoke everything into existence. He is everlasting, unchanging, omnipotent, omniscient, and omnipresent. Those are critical distinctions, and scholars who ignore them are being dishonest or revealing their bias when they dismiss the God of the Bible as the product of theological plagiarism.

And there are more than a few poor souls who believe the Sumerian myths, but with a twist: The Anunnaki, the Mesopotamian gods, were prehistoric astronauts from another world who created mankind from ape DNA. It sounds silly, but that idea has had a good run on a couple of the cable networks in the U.S. and Canada over the last decade. To be blunt, the purveyors of this nonsense either can't read the Sumerian texts, or they can and are deliberately mistranslating them for their own gain.

Yes, another Enemy PSYOP.

The alternate explanation for the similarities between the accounts from the ancient world and those in the Bible is that we're hearing the same story from different points of view: the Bible, which is the true account, and the stories fed to the people in the Fertile Crescent by entities with something to gain by peddling propaganda. It's like the difference in political coverage between competing cable news networks.

Choose wisely which one you believe.

———◆———

The Flood was a cosmic reset for planet earth. The civilization that had developed in what is today southeastern Iraq, a culture based around cities and irrigation agriculture, was forced to start over. You'd think that an event that epic would leave a lasting impression on people. After all, the eight people in the ark had just seen everyone and everything they knew literally washed out of existence.

But no. Within three generations, Yahweh would find it necessary to personally intervene in human affairs again.

The ark likely landed somewhere in what is modern-day far eastern Turkey, near where it borders northwestern Iran and western Armenia. The mountains of Ararat are located within the borders of the ancient nation of Urartu (same name, different language), which covers a lot of the territory occupied by today's Kurds. From there, many of the descendants of Noah, mainly the descendants of Shem and Ham, moved south and east, back down the Tigris and Euphrates to start over.

Interestingly, and probably not coincidentally, scholar Edward Lipinski concluded in the paper we cited earlier that the mountains of Armenia were probably where the abode of El was to be found. Hermon was El's mount of assembly.

Dating the Flood is not an easy task, and there is no date we can present here that will make everybody happy. A more recent date, somewhere around 2300–2200 B.C., will annoy archaeologists and historians, who can point to ruins and tablets that show life continuing pretty much as usual in Mesopotamia throughout the period in spite of evidence of localized flooding here and there. Choosing an earlier date may anger Bible literalists, who prefer to stick with a date calculated by adding the ages of the patriarchs.

Resolving that conflict is a bigger task than we're going to tackle here. By looking at the historical evidence and lining up biblical events with what is attested in Mesopotamian literature and archaeological evidence of big cultural changes, our best guess puts the Flood in the early fourth millennium B.C., somewhere around 3800 B.C. That roughly coincides with the transition from the Ubaid period to the Uruk period. This date isn't likely to please anyone with a passionate belief in a particular year, archaeologist or biblical literalist, but at least it puts the event at a point in history that more or less fits the timeline of the Sumerian King List.

There are still problems with the date of the Flood that won't ever be resolved. For example, Gilgamesh is usually placed in the early third millennium B.C., around 2800 or 2700 B.C. If that's correct, and if there is truth to the Sumerian legend that Gilgamesh visited his ancestor, Utnapishtim (the Mesopotamian Noah), then the timing just doesn't work. It was about a thousand years between the date we've assigned to the Flood and the time of Gilgamesh, and the Bible tells us Noah only (only—ha!) lived another 350 years after the biblical Flood.

And scholars will say they haven't found evidence of a worldwide deluge during that time period. Archaeologists believe the most likely candidates for Noah's Flood were events that left heavy deposits of silt and clay up to eleven feet thick at Ur, Uruk, Kish, and Shuruppak, which is where Utnapishtim—the Akkadian version of Ziusudra, the name recorded in the Sumerian King List—was the last king and lustration priest ("lustration" is the act of making something ritually pure) before "the flood swept over." Unfortunately, those silt layers are dated to different times, ranging from about 3100 B.C. to 2500 B.C. Not only do they contradict one another, the archaeological evidence shows there were obviously survivors who rebuilt and carried on with life at each of those sites as soon as the waters receded.

Our theory, and without hard proof (other than the biblical account) it will never be accepted as more than a theory, is that Noah's descendants returned from the north very soon after the Flood and repopulated the plains of Sumer. Eridu was apparently rebuilt fairly quickly, as were other pre-Flood cities such as Kish, Ur, Uruk, and others.

Skeptics may argue that Mesopotamia of the fourth millennium B.C. was resettled far too quickly to have been depopulated by a global flood. But that's looking backward with a twenty-first-century, Western mindset, one that's based on living in a culture where the average family has 2.1 children and the parents hold off having them until their late twenties or early thirties. That was *not* the perspective of people in the thirty-ninth century B.C., who needed all hands on deck as quickly as possible to build houses, bring in crops, and tend the herds.

Besides, a little math makes it clear that a few centuries were all that was needed to repopulate the Fertile Crescent. In the author's family tree, it was calculated at a family reunion that the patriarch and matriarch of the clan, who married in the 1750s, had produced more than one hundred thousand living descendants by the year 2000. And that's a low number. Our eighteenth- and nineteenth-century ancestors often had six to ten children per family. The last few generations in the twentieth century generally had one or two.

While it's impossible to know who had how many children as the descendants of Noah spread out from the mountains of Ararat (for one thing, the Bible doesn't usually name daughters), it's mathematically possible that his three sons could have produced a couple million descendants within a few hundred years of the Flood.

At some point during this period, we learn of an odd incident between Noah and his youngest son, Ham.

> Noah began to be a man of the soil, and he planted a vineyard. He drank of the wine and became drunk and lay uncovered in his tent. And Ham, the father of Canaan, saw the nakedness of his father and told his two brothers outside. Then Shem and Japheth took a garment, laid it on both their shoulders, and walked backward

and covered the nakedness of their father. Their faces were turned backward, and they did not see their father's nakedness. When Noah awoke from his wine and knew what his youngest son had done to him, he said,

"Cursed be Canaan;
a servant of servants shall he be to his brothers."
He also said,
"Blessed be the LORD, the God of Shem;
and let Canaan be his servant.
May God enlarge Japheth,
and let him dwell in the tents of Shem,
and let Canaan be his servant." (Genesis 9:20–27, ESV)

First, a quick note: Scholars have found that the city of Shuruppak had the biggest granaries in all of Mesopotamia. Thirty-two large, cylindrical, buried silos have been dug up by archaeologists at the site of the ancient city. If Noah actually was Ziusudra/Utnapishtim, the last king of pre-Food Shuruppak, then he came by his knowledge of working the land as keeper of the largest grain stores in the region.

Back to point: Noah's curse seems harsh by our modern standards. Even in the Law given to Moses a couple thousand years later, there is nothing about seeing your father without his clothes. While the actions of Shem and Japheth suggest that the sin of Ham was just voyeurism, what he did was probably much, much worse.

To "see" or "uncover" someone's "nakedness" is an Old Testament euphemism that refers to illicit sexual relations—in this case, incest. Deuteronomy 22:30 and 27:20 are specific prohibitions against a man lying with his father's wife, because that uncovers "his father's nakedness."

There are examples of this among the patriarchs. Reuben went in to Jacob's concubine, Bilhah, the mother of Dan and Naphtali. King David's son, Absalom, after running David and his supporters out of Jerusalem, "went in to his father's concubines in the sight of all Israel" (2 Samuel 16:22).

Why? In that time and place, that was how a son declared himself the new head of the household. Reuben was Jacob's firstborn. Now, it's possible he just found Bilhah attractive and seduced her. But on the other hand, sleeping with Bilhah may have been Reuben's way of claiming his inheritance early. Instead, for desecrating his father's bed, Jacob gave Reuben's share to the sons of Joseph.

Absalom, though, knew *exactly* what he was doing. David's counselor Ahitophel, who had deserted to Absalom's cause, advised the would-be king to pitch a tent on the roof of David's house so all Israel would know that the young man had taken the kingdom. The message to the nation was clear: David was powerless to stop Absalom from doing this outrageous thing and so he was no longer fit to be king.

Absalom's action was, at the risk of being crude, like a dog marking his territory.

Likewise, Ham may have thought that "uncovering his father's nakedness"—having sex with his mother while Noah was too drunk to notice—was a bold move to bypass his brothers and forcibly take a bigger share of inheritance for himself and his children. That could explain why Noah declared the curse on Ham's son, Canaan, rather than on Ham. At that time, the three sons of Noah were basically dividing the world between them.

It may also explain some of the historical-spiritual events that have shaped the world since that day. The descendants of Ham have played a key role in the spiritual and historical events that followed—usually in opposition to the plans of God.

In fact, we might go so far as to say they've been used as a special weapon by the Fallen in their long-running rebellion.

SUMMING UP

Mount Hermon, second on our list of holy mountains, was the site of a key development in the long war by the gods against Yahweh. Some two hundred Watchers, a class of divine being, agreed there to take wives from among human women. The Watchers seem to have traded forbidden knowledge to humanity for access to women, introducing sorcery, astrology, and the arts of war.

The illicit cross-species relations between angels and humans produced the monstrous Nephilim, who threatened to destroy humanity with their violence and insatiable appetites. For their sin, the Watchers were bound with chains in darkness until the judgment. The Nephilim were condemned to die in the Flood and their spirits were doomed to wander the earth as demons until the judgment.

The Watchers were well known to the ancient Mesopotamians as the apkallu. As we're finding with the gods of the ancient world, the apkallu were presented differently in the myths of Sumer, Akkad, and Babylon than in the Bible and other Jewish literature. But there are enough similarities to detect a common origin for the stories—and of course the Enemy uses those similarities to con a skeptical world into believing that the truth is a lie.

History wasn't finished with Mount Hermon just yet. Later, in the historical period, people in the Near East knew it as the mount of assembly for El, the chief god of western Semitic people like the Amorites and Canaanites. It was one of the two main challengers to Yahweh's holy mountain, Zion. And in the chapters ahead, we'll tell you about supernatural battles on the slopes of Hermon that you've probably never heard about in church.

3

BABEL

Not all the holy mountains on our list are natural, formed by the shifting of tectonic plates or the sudden, catastrophic opening of "the fountains of the great deep." The Tower of Babel was one such artificial mountain. Babel was humanity's attempt to force its way back into the divine council.

At Babel, mankind basically tried to storm the castle of God.

For generations, well-meaning Bible teachers have presented the story of Babel as an object lesson on the dangers of pride. Those foolish people were so arrogant they thought they could build a tower high enough to reach heaven!

With all due respect to those teachers, that's an insult to the intelligence of our ancestors, if you think about it. And it's a disservice to people in church who want to know why Yahweh was so offended by this project. *Really? God is that insecure?*

Look, if big egos were enough to bring God to earth, He'd never leave.

Babel was not a matter of God taking down some people who'd gotten too big for their britches. The clue to the sin of Babel is in the name.

Remember, the Hebrew prophets loved to play with language. We often find words in the Bible that sound like the original but make a statement—for example, Beelzebub ("lord of the flies") instead of Beelzebul ("Ba`al the prince"), or Ish-bosheth ("man of a shameful thing") instead of Ishbaal ("man of Ba`al"). Likewise, the original Akkadian words *bāb ilu*, which mean "gate of god" or "gate of the gods," is replaced in the Bible with Babel, which is based on the Hebrew word meaning "confusion."

Now, there's a bit of misinformation that must be corrected about the Tower of Babel: Contrary to what you've heard, Babel was not in Babylon.

It's an easy mistake to make. The names sound alike, and Babylon is easily the most famous city of the ancient world. It's also got a bad reputation, especially to Jews and Christians. Babylon, under the megalomaniacal king Nebuchadnezzar, sacked the Temple in Jerusalem and carried off the hardware for Temple service. It makes sense to assume that a building project so offensive that God personally intervened *must* have been built at Babylon.

But there's a problem with linking Babylon to the Tower of Babel: Babylon didn't exist when the tower was built. It didn't even become a city until about a thousand years after the tower incident, and even then it was an unimportant village for about another five hundred years.

That's a long time. A lot can happen in five hundred years. What was New York City like in 1517? (About a hundred years away from seeing its first European visitors, and about four hundred years from its first internal combustion engine.)

Traditions and sources outside the Bible identify the builder of the tower as the shadowy figure named Nimrod. Our best guess is that he lived sometime between 3500 and 3100 B.C., a period of history called the Uruk Expansion. This tracks with what little the Bible tells us about Nimrod. In Genesis 10:10, we read "the beginning of his kingdom was Babel, Erech, Accad, and Calneh, in the land of Shinar."

The land of Shinar is Sumer and Erech is Uruk. Uruk was so important to human history that Nimrod's homeland is *still* called Uruk, five thousand years later! We just spell it differently—Iraq.

Accad was the capital city of the Akkadians, which still hasn't been found, but was somewhere between Babylon and ancient Assyria. Babylon itself was northwest of Uruk, roughly three hundred miles from the Persian Gulf in what is today central Iraq. But it wasn't founded until around 2300 B.C., at least seven hundred years after Nimrod, and it wasn't really Babylon as we think about it until the old Babylonian empire emerged in the early part of the second millennium B.C.

So where should we look for the Tower of Babel?

Remember, the oldest and largest ziggurat in Mesopotamia was at Eridu, the first city built in Mesopotamia. In recent years, scholars have learned that the name "Babylon" was interchangeable with other city names, including Eridu. So "Babylon" didn't always refer to the city of Babylon in ancient texts. Even though Eridu never dominated the political situation in Sumer after its first two kings, Alulim and Alalgar, the city of Enki was so important to Mesopotamian culture that more than three thousand years later, Hammurabi, the greatest king of the old Babylonian empire, was crowned not in Babylon, but in Eridu—even though Eridu had ceased to be a city about three hundred years earlier.

Even as late as the time of Nebuchadnezzar, 1,100 years after Hammurabi, the kings of Babylon still sometimes called themselves LUGAL.NUN^{ki}—King of Eridu.

Why? What was the deal with Eridu? Yes, it was the first city, the place where "kingship descended from heaven," a city possibly built by Cain or his son, and maybe named for Cain's grandson, Irad.

Think about that for a moment. Eridu—its name interchangeable with Babylon—may have been established by the first murderer on earth. It may have been Cain, not Nimrod, who founded the *original* Babylon!

Archaeologists have uncovered eighteen levels of the temple to Enki at Eridu. The oldest levels of the *E-abzu*, a small structure less than ten feet square, date to the founding of the city around 5400 B.C. Fish bones were scattered around the building. Enki seems to have been a fan of Euphrates River carp.

Now, stop and take that in: The first small shrine to Enki may have been built by Cain or one of his immediate descendants. And consider that the spot remained sacred to Enki long after the city was deserted around 2000 B.C. The temple remained in use until the fifth century B.C., nearly five thousand years after the first crude altar was built to accept offerings of fish to the god of the subterranean aquifer, the *abzu*.

Now, at this point we should tell you that *abzu* (*ab* = water + *zu* = deep) is very likely where we get our English word "abyss."

Ah, the fog lifts!

Another clue: The name Enki is a compound word. *En* is Sumerian for "lord" and *ki* is the word for "earth." Thus, Enki, god of the *abzu*, was "lord of the earth."

Do you remember Jesus calling someone "the ruler of this world"? Or Paul referring to "the god of this world"? Who were they talking about?

Yeah. Satan.

———

Here's another piece to our puzzle: Nimrod was second generation after the Flood. His father was Cush, son of Ham, son of Noah.

In Sumerian history, the second king of Uruk after the flood was named Enmerkar, son of Mesh-ki-ang-gasher.

Enmerkar is also a compound word. The prefix *en* means "lord" and the suffix *kar* is Sumerian for "hunter." So Enmerkar was Enmer the Hunter. Sound familiar?

> Cush fathered Nimrod; he was the first on earth to be a mighty man.
>
> He was a mighty hunter before the LORD. Therefore it is said, "**Like Nimrod a mighty hunter before the LORD.**" (Genesis 10:8–9, ESV, emphasis added)

The Hebrews, doing what they loved to do with language, transformed Enmer—the consonants N-M-R (remember, no vowels in ancient Hebrew)—into Nimrod, which makes it sound like *marad*, the Hebrew word for "rebel."

Now, get this: An epic poem from about 2000 B.C. called *Enmerkar and the Lord of Aratta* preserves the basic details of the Tower of Babel story.

We don't know exactly where Aratta was, but guesses range from northern Iran to Armenia. (Which would be interesting. Not only is Armenia located near the center of an ancient kingdom called Urartu, which may be a cognate for Aratta, it's where Noah landed his boat—the mountains of Ararat. So it's possible Nimrod/Enmerkar was trying to intimidate the people—his cousins, basically—who settled near where his great-grandfather landed the ark. But we just don't know.) Wherever it was, Enmerkar muscled this neighboring kingdom to compel them to send building materials for a couple of projects near and dear to his heart.

Some background: The poem refers to Enmerkar's capital city Uruk as the "great mountain." This is intriguing, since Uruk, like most of Sumer, sits in an alluvial plain where there are precisely no mountains whatsoever. Uruk was home to two of the chief gods of the Sumerian pantheon, Anu, the sky god, and Inanna, his granddaughter, the goddess of war and sex. (And by sex, we mean the carnal, extramarital kind.)

While Anu was pretty much retired (like the later Canaanite god El), having handed over his duties as head of the pantheon to Enlil, Inanna played a very active role in Sumerian society. For example, scholars have translated ritual texts for innkeepers to pray to Innana, asking her to guarantee that their bordellos turn a profit.

Apparently, part of the problem between Enmerkar and the king of Aratta, whose name, we learn from a separate epic, was Ensuhkeshdanna, was a dispute over who was Inanna's favorite. One of the building projects Enmerkar wanted to tackle was a magnificent temple to Inanna, the *E-ana* ("House of Heaven"). He wanted Aratta to supply the raw materials. Apparently, this wasn't only because there isn't much in the way of timber, jewels, or precious metal in the plains of Sumer, but because Enmerkar wanted the lord of Aratta to submit and acknowledge that he was Inanna's chosen one. And so Enmerkar prayed to Inanna:

My sister, let Aratta fashion gold and silver skillfully on my behalf for Unug (Uruk). Let them cut the flawless lapis lazuli from the blocks, let them ...... the translucence of the flawless lapis lazuli .... build a holy mountain in Unug. Let Aratta build a temple brought down from heaven—your place of worship, the Shrine E-ana; let Aratta skillfully fashion the interior of the holy *jipar*, your abode; may I, the radiant youth, may I be embraced there by you. Let Aratta submit beneath the yoke for Unug on my behalf.[7]

Notice that Inanna's temple was, like Uruk, compared to a holy mountain. And given the type of goddess Inanna was, the embrace Enmerkar wanted was more than just—ahem—a figure of speech.

To be honest, some of the messages between Enmerkar and Ensuhkeshdanna about Inanna were the kind of locker room talk that got Donald Trump into trouble during the 2016 presidential campaign. But I digress.

Well…no. Let's continue the digression for a minute. We should stop for a brief look at Inanna's role in human history. The goddess has been known by many names through the ages: Inanna in Sumer, Ishtar in Babylon, Astarte in Canaan, Aphrodite in Greece, and Venus across the Roman world. Let's just say the image we were taught of Aphrodite/Venus in high school mythology class was way off.

Since we'd like to keep this a family-friendly book, we won't dig *too* deeply into the history and characteristics of Inanna. Scholars don't completely agree on the details, anyway. But it's safe to say Inanna wasn't a girl you'd bring home to meet your mother.

In fact, she wasn't always a girl, period. You see, while Inanna was definitely the goddess with the mostest when it came to sex appeal, she was also androgynous. She was sometimes shown with masculine features like a beard. On one tablet (although from much later, in the first millennium B.C., almost three thousand years after Nimrod), Inanna says, "When I sit in the alehouse, I am a woman, and I am an exuberant young man."[8]

Her cult followers included eunuchs and transvestites, and she was apparently the first in history to make a practice of sex reassignment:

> She [changes] the right side (male) into the left side (female),
> She [changes] the left side into the right side,
> She [turns] a man into a woman,
> She [turns] a woman into a man
> She ador[ns] a man as a woman,
> She ador[ns] a woman as a man.[9]

It's wonderfully ironic. The twenty-first-century progressive ideal of gender fluidity was personified more than five thousand years ago by the Sumerian goddess Inanna, a woman who craved sex and fighting as much (or more) than men, taking on all comers in love and war, and better than men at both. Her personality is celebrated by modern scholars as complex and courageous, transcending traditional gender roles, turning Inanna into an icon of independent man/woman/other-hood.

There is an ongoing debate among scholars as to whether the priesthood of Inanna was involved in ritual sex. The concept of divine marriage was common in ancient Mesopotamia, but generally the participants were a god and his consort. It appears that the rituals were intended to please the god so he'd be receptive to the requests from a city or kingdom under his protection.

However, as a *harimtu*, which might mean "temple prostitute" or may simply refer to a single woman, Inanna herself participated in the rite with a king. And since she was the dominant partner in the ritual coupling, gender roles might not have been as clearly defined as we would assume.

From a Christian perspective, however, Inanna isn't complex at all. She's a bad Hollywood screenwriter's idea of a fifteen--year-old boy's fantasy woman. Inanna is selfish, ruled by her passions, and destructive when she doesn't get her way. The Sumerian hero Gilgamesh, who ruled Uruk two generations after Enmerkar, is remembered partly for rejecting Inanna. As he pointed out in the story, every one of the men in her life suffered horrible consequences—for example, Dumuzi the Shepherd, who ruled as a king in Bad-Tibara, the second city in Sumer to exercise kingship after Eridu.

In the myth, even though Inanna married Dumuzi, she was happy to throw him under the bus when demons tried to drag her younger son, Lulal (Bad-Tibara's patron god), down to the netherworld. At Inanna's urging, the demons spared Lulal and took Dumuzi instead. Dumuzi's sister pleaded for him, so Inanna agreed to allow her to take his place for half the year, thus making Dumuzi the first of many "dying and rising gods" in the ancient Near East.

More than two thousand years later, one of the abominations God showed the prophet Ezekiel was women at the entrance of the north gate of the Temple weeping for Dumuzi, called Tammuz in the Bible.

Well, for his impudence at daring to remind Inanna about the fate of Dumuzi, and the other poor shlubs who'd succumbed to the charms of the wild goddess, she flew up to heaven in a rage and demanded that her father, the sky god Anu, unleash the Bull of Heaven on Gilgamesh. That didn't go well for the Bull of Heaven, but sadly for Gilgamesh, his best friend Enkidu was killed by the gods as punishment for spoiling Inanna's revenge.

We shared all of that with you to make a point: *This* is the deity Enmerkar/Nimrod wanted to make the patron goddess of his city, Uruk! (Replacing her father Anu, ironically.) Could it be that veneration of the violent, sex-crazed, gender-bending Inanna was responsible for Yahweh's decision to stop Nimrod's artificial holy mountain?

Well... no. Probably not. Inanna has enjoyed a very long run near the top of the Most Popular Deities list. And why not? Selling humans on the concept of sex as worship is easy.

Looking at the values of our modern society, it's no stretch to say that Inanna is the spirit of the age. Gender fluidity is the flavor of the month among progressives in the West. The values of Inanna—immediate gratification and sex with whoever, whenever—are considered more open-minded, tolerant, and loving than the virtues of chastity, fidelity, and faithfulness introduced by Yahweh long after Inanna was first worshiped as the Queen of Heaven.

Ironically, this means that so-called progressive ideas about gender and sexual morality are actually *re*gressive! The enlightened think they're cutting edge, breaking new ground, and smashing old paradigms, when in fact they're just setting the calendar back to more than a thousand years before Abraham.

If Yahweh had genuinely intervened to put a stop to the cult of Inanna, she would be long forgotten, like Enki. And Miley Cyrus would be a freak, not a culture hero.

No, the transgression of Nimrod was much more serious. Besides building a fabulous temple for the goddess of prostitutes, he also wanted to expand and upgrade the *abzu*—the abyss.

> Let the people of Aratta bring down for me the mountain stones from their mountain, **build the great shrine for me, erect the great abode for me, make the great abode, the abode of the gods**, famous for me, make my *me* prosper in Kulaba, make the *abzu* grow for me like a holy mountain, **make Eridug** (Eridu) **gleam for me like the mountain range, cause the *abzu* shrine to shine forth for me like the silver in the lode.** When in the *abzu* I utter praise, when I bring the *me* from Eridug, when, in lordship, I am adorned with the crown like a purified shrine, when I place on my head the holy crown in Unug Kulaba, then may the ...... of the great shrine bring me into the *jipar*, and may the ...... of the *jipar* bring me into the great shrine. May the people marvel admiringly, and may Utu (the sun god) witness it in joy.[10] (Emphasis added)

That's the issue Yahweh had with it right there. This tower project wasn't about hubris or pride; it was to build the abode of the gods, an artificial mount of assembly, right on top of the abzu.

Could Nimrod have succeeded? Ask yourself: Why did Yahweh find it necessary to personally put a stop to it? A lot of magnificent pagan temples were built in the ancient world, from Mesopotamia to Mesoamerica. Why did God stop this one?

We can only speculate, of course, but there's a reason or we wouldn't have a record of it. In fact, the fact that it's in the Bible at all means it's an important enough lesson that God wanted to preserve it for us. Calling Babel a sin of pride is easy, but it drains the story of its spiritual and supernatural context. And frankly, it makes God seem a bit insecure. Doesn't He want us to work hard to reach our goals?

Well, Inanna gave Enmerkar the answer he wanted (said Enmerkar, anyway; history is always written by the winner), so the king chose a messenger to carry this message to the lord of Aratta:

> Lest I make the people fly off from that city like a wild dove from its tree, lest I make them fly around like a bird over its well-founded nest, lest I requite (?) them as if at a current market rate, lest I make it gather dust like an utterly destroyed city, lest like a settlement cursed by Enki and utterly destroyed, I too utterly destroy Aratta; lest like the devastation which swept destructively, and in whose wake Inanna arose, shrieked and yelled aloud, I too wreak a sweeping devastation there—let Aratta pack nuggets of gold in leather sacks, placing alongside it the *kugmea* ore; package up precious metals, and load the packs on the donkeys of the mountains.[11]

Enmerkar justified his demand and his threat by claiming that Inanna had chosen him and Uruk as her favorites and sent his messenger off across "seven mountains" to reach Aratta. That's possibly coincidental, but an interesting reference nonetheless considering end-times prophecies about the seven mountains of Babylon the Great (see Revelation 17:1–14, especially verse 9). Unintimidated, the lord of Aratta, Ensuhkeshdanna, refused to submit. Instead, he proposed a series of challenges that he believed were impossible to fulfill, but Enmerkar, with the help of Enki, succeeded anyway.

There's more to this story, which was apparently so popular in Mesopotamia even as late as the old Babylonian era (early third millennium B.C., more than 1,500 years later), that several other surviving Sumerian epics read like sequels to a hit movie. Without getting bogged down in details, we can draw a few broad conclusions from the tale: The king of Uruk, whom we believe was Nimrod, wanted to build a fabulous temple to the goddess of sex and war as a centerpiece of his capital city. More important, Enmerkar/Nimrod wanted to rebuild and expand the ancient, pre-flood temple of Enki, the god Sumerians thanked for the gifts of civilization (the mes) and for creating the apkallu (the Watchers)—even though their intervention in human affairs, according to later Babylonian myth, was responsible for the flood.

But there's another fascinating detail recorded in *Enmerkar and the Lord of Aratta*: In the story, Enki confused the speech of humans, who had formerly spoken the same language.

Once upon a time there was no snake, there was no scorpion,
    There was no hyena, there was no lion,
    There was no wild dog, no wolf,
    There was no fear, no terror,
    Man had no rival.
    In those days, the lands of Subur (and) Hamazi,
    Harmony-tongued Sumer, the great land of the decrees of princeship,
    Uri, the land having all that is appropriate,
    The land Martu, resting in security,
    The whole universe, the people in unison
    To Enlil in one tongue [spoke].
    (Then) Enki, the lord of abundance (whose) commands are trustworthy,
    The lord of wisdom, who understands the land,
    The leader of the gods,
    Endowed with wisdom, the lord of Eridu

Changed the speech in their mouths, [brought] contention into it,

Into the speech of man that (until then) had been one.[12]
(Emphasis added)

The elements of the Babel account are all there in *Enmerkar and the Lord of Aratta*: The pride that drove Nimrod to dominate the known world, construction of a tower/ziggurat that Yahweh found offensive enough to bring to a halt (building the abode of the gods over the abyss!), and the supernatural confusion of languages that gave the tower its Hebrew name, Babel.

In the view of this author, the evidence is compelling. It's time to correct the history we've been taught since Sunday School: Babel was not at Babylon, it was at Eridu. The tower was the temple of the god Enki, Lord of the Earth, the god of the abyss. Its purpose was to create an artificial mount of assembly, the abode of the gods, to which humans had access.

That was something that Yahweh could not allow.

The end of *Enmerkar and the Lord of Aratta* is mostly missing, but it appears that Enmerkar ultimately triumphed over his rival. Other stories suggest that Enmerkar later marched the army of Uruk to Aratta and conquered it.

This is consistent with archaeological evidence of the Uruk Expansion, which covers the period from about 3500 B.C. to about 3100 B.C. Although scholars usually downplay the violence that created the world's first empire, Uruk spread its influence as far away as northwest Iran and southeastern Turkey. Pottery from Uruk has been found more than five hundred miles away from the city. To put it into context, Uruk at its peak controlled more territory than Iraq under Saddam Hussein.

This was not always a peaceful endeavor. An ancient city called Hamoukar in northeast Syria was destroyed and burned by an army from Uruk sometime around 3500 B.C. Scholars have identified the origin of the army by the pottery they left behind. Hamoukar was overwhelmed and then burned by attackers who used clay bullets fired from slings to defeat the city's defenders. Strangely, what appears to have been a trading post from Uruk outside the city was destroyed, too, suggesting that maybe the men sent by Uruk to keep the locals in line had gone native.

That was how the kingdom of Nimrod obtained materials like jewels, copper, silver, lead, gold, timber, wine, and other things that were scarce in the plains of Sumer.

Of course, there is no way we'll ever know for certain that Nimrod was Enmerkar, and that he was responsible for the Uruk Expansion—which is a nice way of describing the process of conquering everybody within a two-month march of home. Artifacts from Uruk are found everywhere in the Near East, especially a type of pottery called the beveled-rim bowl. This is significant because it offers a glimpse into the way the society of Uruk was organized.

We described earlier how the society just before the Uruk period, the Ubaid culture, became more stratified as people moved from rural settlements to cities. The Ubaid civilization produced high-quality pottery, identified by black geometric designs on buff or green-colored ceramic. In contrast, around 3500 B.C., the Uruk culture developed the world's first mass-produced product, the beveled-rim bowl.

The beveled-rim bowl is crude compared to the pottery from the Ubaid culture, but archaeologists have found a *lot* of them. About three-quarters of all pottery found at Uruk period sites are beveled-rim bowls. Scholars agree that these simple, undecorated bowls were made in molds rather than on wheels, and that they were probably used to measure out barley and oil for workers' rations.

The way they were produced left the hardened clay too porous to hold liquids like water or beer. (Yes, the Sumerians brewed beer. Enki's alternate name, Nudimmud, is a compound word: *nu* = "likeness" + *dim* = "make" + *mud* = "beer." One could argue that Enki [Nudimmud], like Inanna, is a spirit of our age.) The bowls were cheap and easy to make, so much so that they may have been disposable. At some archaeological sites, large numbers of used, unbroken bowls have been found in big piles.

Basically, these cheap bowls were the Sumerian version of Styrofoam fast-food containers.

The concept of measuring out rations implies an employer or controlling central authority responsible for doling out grain and oil to laborers. It's not a coincidence that the development of these crude bowls happened alongside Uruk's emergence as an empire. After the Flood, which we theorize marked the end of the Ubaid period, people again gravitated to urban settlements where they apparently exchanged their freedom for government rations.

It looks like that's how Nimrod and his successors, including Gilgamesh, controlled their subjects—move them off the land and into cities, keeping a tight rein on the means of production and distribution of food and resources.

At the risk of getting repetitious, things haven't changed all that much over the last five thousand years. But I digress.

Now, it's possible we're reading more into the evidence than is truly there. It could be that the beveled-rim bowl was nothing more than an easy way for people to carry lunch to work. Will future archaeologists conclude that Americans were paid in McBurgers because of the billions of Styrofoam containers in our landfills?

Still, given the unprecedented growth of the Uruk empire between about 3500 B.C. and 3100 B.C., it's not going too far to speculate that the use of mass-produced ration bowls was a symptom of the stratification of society under the rule of Enmerkar and his successors. As in the Ubaid culture, citizens of Uruk found themselves working for hereditary leaders—kings, who justified their rule as ordained by the gods.

As an example, the Sumerian myth *Enki and Inanna* tells the story of how the divine gifts of civilization, the *mes*, were stolen from Enki by Inanna and transferred from Eridu to Uruk. Enki, always ready for a romp with a goddess, tried to ply Inanna with beer. She maintained her virtue while Enki got drunk, offering her gift after gift as his heart grew merry and his mind grew dim. When he awoke the next day with a hangover, Inanna and the mes were no longer in the *abzu*. The enraged god sent out his horrible *gallu* demons (sometimes translated "sea monsters") to retrieve them, but Inanna escaped and arrived safely back at Uruk, where she dispensed the hundred or so mes to the cheers of a grateful city. Enki realized he'd been duped and accepted a treaty of everlasting peace with Uruk. This tale may be a bit of religious propaganda to justify the transfer of political authority from Eridu to Uruk.

One more thing: We mentioned earlier that archaeologists at Eridu have found eighteen construction layers at the site of Enki's temple. Some of those layers are below an eight-foot deposit of silt from a massive flood. The most impressive layer of construction, called Temple 1, was huge, a temple on an enormous platform with evidence of an even larger foundation that would have risen up to almost the height of the temple itself.

Here's the thing: Temple 1 was never finished. At the peak of the builders' architectural achievement, *Eridu was suddenly and completely abandoned.*

> The Uruk Period…appears to have been brought to a conclusion by no less an event than **the total abandonment of the site**…. In what appears to have been an almost incredibly short time, drifting sand had filled the deserted buildings of the temple-complex and obliterated all traces of the once prosperous little community.[13] (Emphasis added)

Why? What would possibly cause people who'd committed to building the largest ziggurat in Mesopotamia at the most ancient and important religious site in the known world to just stop work and leave Eridu with the *E-abzu* unfinished? Could it be…

> "Come, let us go down and there confuse their language, so that they may not understand one another's speech." So the LORD dispersed them from there over the face of all the earth, **and they left off building the city.** (Genesis 11:8–9, ESV, emphasis added)

Contrary to the Sumerian record, it was Yahweh who mixed the speech of the Sumerians "that (until then) had been one." Claiming this feat for Enki was another PSYOP by the Fallen.

To the Sumerians, and later the Akkadians and Babylonians (who knew him as Ea), Enki was the supernatural actor with the most influence on human history. He was the caretaker of the divine gifts of civilization, the mes (at least until he was tricked by Inanna), and he retained enough prestige for powerful men to justify their reign by claiming kingship over his city, Eridu, for 2,500 years after the city was abandoned and became nothing but a temple complex.

The consequences of Babel were immediate and severe. Not only did Yahweh scatter the people, He told mankind that we'd have to deal with the lesser *elohim*—the small-G gods—from then on.

It's not immediately obvious in the Genesis account, but Moses reminded the Israelites of what went down in the aftermath of Yahweh's divine intervention at Babel:

> When the Most High gave to the nations their inheritance,
> **when he divided mankind,**
> **he fixed the borders of the peoples**
> **according to the number of the sons of God.**
> **But the LORD's portion is his people,**
> Jacob his allotted heritage.
> (Deuteronomy 32:8, ESV, emphasis added)

Most English translations render the last words of verse 8 "sons of Israel". The English Standard Version translators follow most of the existing copies of the Septuagint and the texts of Deuteronomy found among the Dead Sea Scrolls. But even just on a logical basis, "sons of God" makes more sense. Israel (Jacob) wasn't around when God divided mankind after Babel. He wouldn't be born for another 1,200 to 1,500 years.

This is important. Getting this verse right makes a lot of things in the Bible easier to understand. Making the case linguistically requires going back to a couple of earlier passages in Deuteronomy.

> And beware lest you raise your eyes to heaven, and when you see the sun and the moon and the stars, all the host of heaven, you be drawn away and bow down to them and serve them, **things that the** LORD **your God has allotted to all the peoples under the whole heaven.** (Deuteronomy 4:19, ESV, emphasis added)

> All the nations will say, "Why has the LORD done thus to this land? What caused the heat of this great anger?" Then people will say, "It is because they abandoned the covenant of the LORD, the God of their fathers, which he made with them when he brought them out of the land of Egypt, and went and served other gods and worshiped them, **gods whom they had not known and whom he had not allotted to them.**" (Deuteronomy 29:24–26, ESV, emphasis added)

The word rendered "inheritance" and "heritage" in Deuteronomy 32:8–9 is the Hebrew word *nakhal,* and the word translated "portion" is the Hebrew *kheleq.* In the Deuteronomy 4 and 29 passages, "allotted" is based on the Hebrew *khalaq,* the same root used to describe Israel's status in chapter 32. Israel is the allotted heritage of Yahweh, and the other gods were allotted to "all the peoples under the whole heaven." So after Babel, the nations of the world were divided, geographically and spiritually, according to the number of the sons of God.

The concept of territorial spirits is biblical, although it's not often mentioned in Sunday morning services. The most obvious example is in a previously cited passage from the book of Daniel:

I lifted up my eyes and looked, and behold, a man clothed in linen, with a belt of fine gold from Uphaz around his waist....

Then he said to me, "Fear not, Daniel, for from the first day that you set your heart to understand and humbled yourself before your God, your words have been heard, and I have come because of your words. **The prince of the kingdom of Persia withstood me twenty-one days, but Michael, one of the chief princes, came to help me, for I was left there with the kings of Persia**, and came to make you understand what is to happen to your people in the latter days. For the vision is for days yet to come."...

But now I will return to fight against the prince of Persia; and when I go out, behold, the prince of Greece will come. But I will tell you what is inscribed in the book of truth: there is none who contends by my side against these except Michael, your prince." (Daniel 10:5, 12–13, 20–21, ESV, emphasis added)

There are a couple of intriguing questions here: Who was the man Daniel saw? And who is the Prince of Persia? Unfortunately, the answers aren't nearly as interesting because we just don't know for sure.

Some scholars think Daniel saw an angel of high rank, while others note similarities between the description of the angel and John's description of Jesus in the Book of Revelation. However, Jesus wouldn't need help dealing with another spirit being, so Daniel's visitor must have been an angel. It may have been Gabriel, who Daniel saw in the two previous chapters, but then why didn't he just identify the man since he'd seen him at least twice before? The bottom line: An angel, possibly one not named in Scripture.

The Prince of Persia was certainly a spiritual entity because a human opponent wouldn't have been able to hold him at all, much less for three weeks! But identifying that entity is an impossible task. We can speculate, though, that it was one of the seventy *bene elohim* who were allotted to the nations.

How do we know there were seventy? Good question. The nations are named in Genesis chapter 10, the section called the "Table of Nations."

Quick quiz: How many nations are listed in Genesis chapter 10? Right, seventy.

Now, go back to our earlier chapter on Mount Hermon if you need to look this up: How many sons of El gathered in the assembly on Mount Hermon?

Seventy.

Is this a coincidence? *No!*

These seventy bene elohim apparently presented themselves to the ancient world as their gods. The aftermath of Babel is where the concepts of holy ground and national gods began. It was Dagon for the Philistines, Chemosh for Moab, Molech for Ammon, Marduk for Babylon, Hadad (Ba`al) for the Canaanites, and so on. "But the LORD's portion is his people, Jacob his allotted heritage" (Deuteronomy 32:9).

The Mesopotamians were already heading down that path when God made the arrangement formal. Every city in Sumer had a temple that was home to the city's patron god. This sometimes led to war as city-kings went out to conquer on orders of their god. (To be fair, that claim has been made by alleged Christians from time to time.) The political situation in Sumer and the greater ancient Near East often reflected the relative position of various gods within the pantheon.

So how did the rebellion work out for those seventy sons of God? About as well as it did for the Watchers on Mount Hermon. Because the bene elohim rebelled and began accepting worship, Yahweh passed judgment on them, too. Refer back to Psalm 82. The gods are under a death sentence, and a day is coming when God inherits all the nations!

SUMMING UP

Contrary to what we've been taught, the Tower of Babel was not the ziggurat devoted to Marduk at Babylon. That was still a thousand years in the future when Yahweh descended from heaven to confuse the speech of Babel's work crews.

Nimrod, the Sumerian king Enmerkar, tried to rebuild and expand an ancient temple at the city of Eridu dedicated to Enki, the god of the abyss. His purpose was to create an artificial "abode of the gods." The Tower of Babel, the *bab ilu* or "gate of the gods," was Enki's *E-abzu*—the House of the Abyss.

Yahweh personally intervened to stop the work by confusing the languages of the people. Because of their rebellion, He divided the people into seventy nations, establishing their borders according to the number of the sons of God, the bene elohim. He then allotted the bene elohim to the nations as divine overseers. But they set themselves up as gods and failed to uphold Yahweh's standards of justice. And so Yahweh held judgment in the divine council. He decreed that the bene elohim would one day die like men and He would inherit all the nations.

But the bene elohim didn't give up without a fight. And the Bible records their rebellion in living color—if you know what to look for.

4

SINAI

LET'S FAST FORWARD about a thousand years from Babel. After the tower was abandoned, it appears that a group of Sumerians traveled by sea around the Arabian Peninsula, and then overland across the wadis extending west from the Red Sea to found the first dynasty of Egypt.

This isn't as crazy as it sounds. Early Egyptologist William Matthew Flinders Petrie noted a sharp difference between two groups of people buried at a large site near the village of Nakada in Upper (southern) Egypt. One group had been interred with very basic grave goods in simple pits covered with palm branches. The second group had been ritually dismembered, buried in pits lined with brick along with objects of value, such as lapis lazuli jewelry, and then covered with palm logs.

Petrie eventually theorized that the second group, which he dubbed the Falcon Tribe, had invaded and conquered the native inhabitants with superior technology, such as the pear-shaped mace found buried with some in the second group. Make no mistake, in the fourth millennium B.C., the pear-shaped mace was a weapon of mass destruction.

Other evidence, from artwork to architecture—for example, Egypt's first pyramid, for the pharaoh Djoser, is clearly modeled on the Sumerian ziggurat—linked the so-called Dynastic Race with Mesopotamia. This theory was widely accepted until World War II. After Hitler, however, the Dynastic Race concept was a little too much like the Nazis' ideas about genetics and bloodlines for comfort.

But then in 1995, Egyptologist David Rohl published his first book, *A Test of Time*. Rohl makes a strong case for the Dynastic Race theory, even documenting ancient graffiti in Egypt that appeared to show the Falcon Tribe carrying their boats overland from the Red Sea toward the Nile.

Now, is it a coincidence that the name of the first king of the first Egyptian dynasty, Narmer, is awfully close to that of Nimrod, the would-be emperor of Uruk? Scholars have to guess at vocalization in many cases. It's not too much of a stretch to suggest that the names Narmer and Enmerkar (or "Enmer the Hunter") were the same.

Peter D. Goodgame explored this idea in greater depth in his book *Second Coming of the Antichrist*. I bring it up here only to suggest one possible explanation for the decline of Uruk as a regional power at just about the time of the Tower of Babel on the timeline of history. After the humiliation at Babel, Nimrod/Enmerkar may have decided to head for new lands and a fresh start. Since his father, Cush, was apparently the founder of Ethiopia, nearby Egypt, which had been settled by Cush's brother, Mizraim, may have been a logical place to start over.

Think about that. Did Cush throw his brother and his brother's family under the bus to give his son, Nimrod, a new start?

By the way, the Sumerian King List mentions that the first city to receive the kingship after the Flood was Kish. Remember, there are no vowels in ancient Hebrew. Could Kish have been named for Cush? And Enmerkar's father, Mesh-ki-ang-gasher, is said to have "entered the sea and disappeared." Maybe that's how Sumerian scribes remembered King Cush, who sailed off into the Persian Gulf to establish a new colony with his brothers Mizraim (Egypt) and Put (Libya) in Africa.

Now, that's all speculation. Nothing so far in this chapter can be established for certain. And to be honest, the Sumerian King List has way too many names between the founding of the Kish dynasty and Enmerkar for the Kish/Cush theory to work.

As the first dynasties in Egypt established themselves and began to build monuments that would surpass those left behind in Sumer, empires rose and fell between the great rivers Euphrates and Tigris. Akkad, under Sargon the Great, established a kingdom around 2350 B.C. that stretched from the Persian Gulf almost all the way to the Mediterranean, but it collapsed less than two hundred years later under the weight of invasions from the barbaric Guti, who swept onto the plains of Sumer from the Zagros Mountains in northwest Iran.

The Guti, about whom we know very little because they didn't write, controlled Mesopotamia for about fifty years. They were finally thrown out by Utu-hengal, a king of Uruk, which set off a struggle for dominance between the city-states of the region. Ur finally emerged supreme, and what scholars call the Third Dynasty of Ur gave the region its last native Sumerian kings for a brief period, until about 2000 B.C. Then Ur was sacked by its ancient rival, the Elamites, who occupied what is now the far west and southwest of Iran, the region along the east side of the Persian Gulf.

Into that power vacuum moved a group of Semitic-speaking people called the Amorites. Scholars think the Amorites originated in central Syria, around a mountain called Jebel Bishri, which is on the west side of the Euphrates between Deir ez-Zor and Raqqa. However, scholars still debate that point more than a century after the first scholarly books were published about the Amorites.

Academics aren't even sure the Amorites conquered the various existing Akkadian and Sumerian city-states. They may have been part of Mesopotamian culture all along and just somehow came out of the confusion around the turn of the millennium in control of the political machinery. What we do know is that around the beginning of the twentieth century B.C., Amorite kingdoms emerged in what had formerly been Subartu (Assyria), Akkad, and Sumer, and, along with Amorite kingdoms in the Levant, they dominated the Fertile Crescent for the next four centuries.

This was the world of Abram. We'll have more to say about the Amorites, but know this: Their influence on history is much greater than you've been told.

Here's another bit of inaccurate history we've been taught: Abram, later Abraham, didn't come from the Ur in southeastern Iraq, the one that was in imminent danger of being torched by the Elamites. Although it seems to make sense that he might have been a refugee from the collapse of Ur, it's far more likely that Abraham was born and raised in a part of the world that was close to the Amorite heartland, near the border between modern-day Syria and Turkey.

This was the belief of most scholars for many years until famed archaeologist Sir Leonard Woolley made his spectacular discoveries at Ur in Iraq. That Ur, with its magnificent ziggurat and stunning "royal tombs," seemed much more appropriate as the ancestral home of the patriarch of the world's great monotheistic religions than someplace in Turkey that hasn't been found yet.

A recent discovery about some of the remains found by Woolley at Ur gives us a glimpse into the world of Abram. It also illustrates a tendency among scholars who study ancient civilizations to view the ancient world through rose-colored glasses. Woolley and his team, who worked at Ur in the '20s and early '30s, found 1,850 burials dated to the second half of the third millennium B.C. Seventeen were so elaborate that Woolley, displaying a flair for marketing, dubbed them the "Royal Tombs of Ur."

One tomb in particular is worth our attention. Somehow, it escaped being plundered by tomb raiders over the years. It was the tomb of a noblewoman by the name of Pu'abi, an Akkadian name that means "commander of the father." She's believed to have died around 2600 B.C. Pu'abi was buried wearing a fabulous golden headdress adorned with carnelian and lapis lazuli. Lapis lazuli wasn't easy to get back in the day; it was only found in Afghanistan, which is still a major source of the semiprecious blue stone, and shipped to Sumer by way of Meluhha, a civilization on the west coast of India.

Among the other treasures buried with Pu'abi, Woolley found the famous Golden Lyre of Ur, one of a dozen stringed instruments in the tomb when it was opened in 1929. Sadly, the Golden Lyre of Ur is one of the priceless treasures of antiquity that was lost when the Baghdad Museum was looted in 2003. It was found in pieces in the museum's car park.

But more to the point for our purposes: Also buried with Pu'abi were fifty-two other people arranged in rows inside her tomb. These were apparently servants sent to the afterlife with her to ensure that Pu'abi had everything she needed for eternity.

How nice for her.

Now, Woolley, interpreting the scene with a romantic bias, decided that the servants had gone to their eternal rest willingly, drinking some toxic elixir and then peacefully lying down to await whatever came next. But in 2011, researchers from the University of Pennsylvania used CT scanners to examine six skulls from different royal tombs, and they reached a much darker conclusion.

Digital imaging technology and modern forensic science made it clear that the actual cause of death in all six cases was blunt force trauma. Instead of quietly drifting off to their eternal rest, the victims had been bashed in the back of the head with the business end of a battle-axe.

That dumped a large bucket of ice water onto Woolley's vision of an idyllic death scene. Keep that in mind the next time you get really aggravated at the sense of entitlement displayed by politicians, pop divas, and first-round draft choices. At least when *they* die, they don't take dozens of people with them.

This was the world into which the patriarchs were born.

Well, because of Woolley's truly incredible discoveries, Jews and Christians revised the standard map of Abraham's journeys to show a long trek from southeast Iraq to southeast Turkey, which, ironically, is roughly the route the army of Uruk took on its way to destroying Hamoukar about 1,500 years earlier.

Prior to Woolley, scholars assumed that Abraham's Ur was somewhere near Harran, about ten miles inside Turkey along the Balikh River, a tributary of the Euphrates that joins the great river at the Syrian city of Raqqa. Harran was a merchant outpost in Abraham's day, and it was perfectly situated for it. It sat on a trade route from the Mediterranean to Sumer, linking the cities of Antioch and Carcemish with Nineveh, Babylon, and beyond.

This is consistent with what we know of Abram's livelihood, who seems less like a shepherd and more like a traveling merchant who conducted business with men at the highest levels of government, as evidenced by his interactions with local kings in Canaan and the pharaoh in Egypt.

Harran is also the name of one of Abraham's brothers. It's probable that his brother was named for the city and not the other way around, since the city first appears among records recovered from the ancient city of Ebla dated to about 2300 B.C., some three hundred years before Abraham.

The great scholar of Near Eastern history and ancient languages Cyrus H. Gordon made a strong case for Abraham's origins along the upper Euphrates rather than in the heart of Sumer. Gordon, writing in the *Journal of Near Eastern Studies* in 1958,[14] highlighted a then-recently translated Akkadian tablet from the ancient Canaanite city of Ugarit, a decree from mid-thirteenth century B.C. by the powerful Hittite king Hattusili III to the king of Ugarit, Niqmepa. This decree regulated the activity of Hittite merchants operating in Ugarit, and identified the merchants in question as citizens of Ura, a city near Harran that specialized in *tamkârûtum*, or foreign trade.

Niqmepa had apparently complained to Hattusili about the traders from Ura. In response, Hattusili decreed that the merchants could conduct business in Ugarit only during the summer months (no great loss—that was the only time of the year when farmers had crops with which to pay);

the merchants would have to return home to Ura in the winter; and they were barred from buying real estate in Ugarit.

Gordon then pointed out that the Genesis account indicates that while Abraham and his descendants were in Canaan at God's call, they were also there for the express purpose of conducting trade. For example:

> But Hamor [king of Shechem] spoke with [Jacob and his sons], saying, "The soul of my son Shechem longs for your daughter. Please give her to him to be his wife. Make marriages with us. Give your daughters to us, and take our daughters for yourselves. You shall dwell with us, and the land shall be open to you. **Dwell and trade in it, and get property in it.**" (Genesis 34:8–10, ESV, emphasis added)

Gordon observed that men who traded abroad, like Abraham, Isaac, and Jacob, apparently looked for opportunities to settle where they did business, but were generally barred from buying real estate in foreign lands.

So there are enough similarities between the lifestyles of the patriarchs and the merchants of Ura, not to mention Ura's proximity to Harran, to make the identification of Ura as Abraham's Ur a strong possibility. Gordon went on in the article to demonstrate the linguistic possibility that the final vowel in Ura was dropped in the transition from Aramaic to Hebrew.

Logistically, Harran was way too far north and east for Abraham's father Terah to have made it a stop on the way to Canaan if he'd been traveling from Ur in Sumer. There were much shorter routes between Sumer and Canaan—for example, a trade route used by Amorites that linked Mari on the Euphrates to Damascus via Tadmor (Palmyra).

In other words, Mesopotamians in the twenty-first century B.C. would have mocked anyone from Sumerian Ur who tried to get to Canaan by way of Harran. Ending up at Harran would have required missing the turn completely and then taking another one in the wrong direction, *away* from Canaan.

Put it this way: Traveling to Canaan from Ur by way of Harran is like driving from Nashville to Kansas City by way of Minneapolis. And a mistake like that is a lot harder to make when you're traveling at the speed of camel.

Finally, Woolley's Ur was on the wrong side of the Euphrates.

> Joshua gathered all the tribes of Israel to Shechem and summoned the elders, the heads, the judges, and the officers of Israel. And they presented themselves before God. And Joshua said to all the people, "Thus says the LORD, the God of Israel, 'Long ago, **your fathers lived beyond the Euphrates,** Terah, the father of Abraham and of Nahor; and they served other gods. **Then I took your father Abraham from beyond the River** and led him through all the land of Canaan, and made his offspring many. I gave him Isaac.'" (Joshua 24:2–3, ESV, emphasis added)

Ur in Sumer was not beyond the Euphrates. It sat on the west bank of the river. But Ura, in modern-day Turkey, was, in fact, "beyond the River."

"But what about those Chaldeans?" you ask. "Weren't they around Babylon?"

Glad you asked.

The Chaldeans were a Semitic tribe, possibly descendants of the Amorites, who founded the Neo-Babylonian empire in the first millennium B.C. That's the Babylon of Nebuchadnezzar, not the Babylon of Hammurabi, who lived about 1,200 years earlier.

But the Greek historian Xenophon, who wrote in the fourth century B.C., referred to another group of Chaldeans. They were a warlike people who were neighbors of the Armenians. They lived north of Ura and Harran, between Mesopotamia and modern Armenia. In another book, Xenophon linked the Chaldeans to the Carduchians, who were probably the ancestors of the modern Kurds. The Kurds today occupy a big swath of territory along the borders between Turkey, Syria, Iran, and Iraq, an area that includes the places we're looking at, Harran and Ura. (And they've been fighting for their own piece of independent real estate for at least 2,400 years. But I digress.)

Gordon also noted that greater Armenia, which was called Urartu from the time of David until the rise of the Neo-Babylonian empire, was in its earliest days known as Ḥaldi (the "ḫ" sounds like "k", so, KALL-dee). In fact, the chief god of the Urartians was named Ḥaldi, father of the Urartian storm-god Teisheba (another manifestation of Ba`al-Hadad).

So it's probable that Ur of the Chaldees was actually Ura of the Ḥaldis. Abraham came from a border region located between the Hittites, the Hurrians, Semitic kingdoms such as Ebla and Mari, Amorite territory, and the Subartans (Assyrians), not the heart of Sumer as we've been taught. Understanding Abraham's origins helps understand the early history of Israel.

In a nutshell, Yahweh needed to give the future Israel time to develop its own identity. That wasn't likely to happen with His people surrounded by the temptations of decadent urban life. Witness what happened to Abraham's nephew, Lot, when he settled in Sodom.

Further, there is a cryptic comment God made to Abraham that relates to the spiritual history of the world down to the present day.

> And he said to him, "I am the LORD who brought you out from Ur of the Chaldeans to give you this land to possess." But he said, "O Lord GOD, how am I to know that I shall possess it?"
>
> He said to him, "Bring me a heifer three years old, a female goat three years old, a ram three years old, a turtledove, and a young pigeon." And he brought him all these, cut them in half, and laid each half over against the other. But he did not cut the birds in half. And when birds of prey came down on the carcasses, Abram drove them away.
>
> As the sun was going down, a deep sleep fell on Abram. And behold, dreadful and great darkness fell upon him. Then the

LORD said to Abram, "Know for certain that your offspring will be sojourners in a land that is not theirs and will be servants there, and they will be afflicted for four hundred years. But I will bring judgment on the nation that they serve, and afterward they shall come out with great possessions. As for yourself, you shall go to your fathers in peace; you shall be buried in a good old age. And they shall come back here in the fourth generation, **for the iniquity of the Amorites is not yet complete.**" (Genesis 15:7–16, ESV, emphasis added)

Iniquity of the Amorites? Wait—what?
Patience. Let's skip ahead to Moses.

———

By the time Moses arrived on the scene, around 1500 B.C., the Hebrews had been in Egypt for more than a hundred years. The days of Joseph serving as vizier to the pharaoh were long gone. The Hebrews had grown from an extended family of about six dozen to a couple million, but they were suffering under the rule of a nation that no longer valued their presence, except as forced labor.

So Yahweh set the next phase of His plan in motion. After guiding the life of Moses from infancy to adulthood (you don't think he survived that trip in the reed boat by accident, do you?), Yahweh appeared to Moses in his exile and tasked him with bringing Israel out of Egypt. And the way God had him do it was a clear message to the gods of Egypt.

Moses' first encounter with Yahweh was in Midian. That was at Horeb in the northern Sinai, later part of Edom (contrary to long tradition that puts the mountain in southern Sinai), the *har elohim*, or mountain of God.

And the angel of the LORD appeared to him in a flame of fire out of the midst of a bush. He looked, and behold, the bush was burning, yet it was not consumed. And Moses said, "I will turn aside to see this great sight, why the bush is not burned." When the LORD saw that he turned aside to see, God called to him out of the bush, "Moses, Moses!" And he said, "Here I am." Then he said, "Do not come near; take your sandals off your feet, for the place on which you are standing is holy ground." And he said, "I am the God of your father, the God of Abraham, the God of Isaac, and the God of Jacob." And Moses hid his face, for he was afraid to look at God. (Exodus 3:2–6, ESV)

Who was this *malak yhwh*, the angel of Yahweh? He appeared to Moses in the burning bush, and it's not clear from the text whether Moses saw the angel, Yahweh, or both. That's not sloppy writing, an error by Moses when he wrote Exodus, or a mistake by a scribe or monk copying a manuscript at some point during the last 3,500 years. This juxtaposition of Yahweh and the angel of Yahweh occurs at other places in the Old Testament.

We don't have the time or space here to unpack it thoroughly in this book. Besides, it's already been done by an actual scholar. At the risk of repetition, please do yourself a favor and get a copy of *The Unseen Realm* by Dr. Michael S. Heiser. It will change your understanding of the Bible (in a good way). The bottom line is that the angel of Yahweh *is* Yahweh, the second power in heaven, and this appearance of the angel is a Christophany—Jesus Christ in the Old Testament.

It's also important to note that Yahweh's first meeting with Moses took place on Horeb, the mountain of God. Get this: The burning bush incident was the first time since Eden that a human had come face to face with Yahweh on His holy mountain. There is no question that the bene elohim, the Fallen, the seventy rebel angels God allotted to the nations after Babel knew about this meeting.

This was a very clear message from Yahweh to the rebels: *I have reestablished my mount of assembly on the earth.*

And they hadn't seen anything yet. Yahweh had an even more spectacular message to deliver through Moses. But let's back up a step before we get to the good part.

Based on the Bible's chronology, Israel was in Egypt for about 215 years, dating from Jacob's arrival to the Exodus. (The 430 years mentioned in Exodus 12:40 is best explained by Moses beginning the count with Abraham's arrival in Canaan. Abraham was, after all, as much the founder of Israel as Jacob, and Canaan was under Egyptian control for most of the second millennium B.C.)

By the time Moses arrived on the scene, Israel had been in the land for more than 130 years. Most of us think of the Israelites' time in Egypt as a horrible experience of backbreaking labor, but it probably wasn't like that for the first seventy-five years or so. According to the Bible's chronology, only sixty-four years passed between the death of Joseph and the birth of Moses, and we know that Moses was eighty when he led the Israelites out of Egypt. So something happened between Joseph's death and the birth of Moses that turned things sour for the house of Israel. But it was necessary for Israel to endure its time in Egypt to build up enough numbers to prevent being absorbed into the surrounding culture in Canaan.

There were hints that this was beginning before Jacob and his sons followed Joseph to Egypt. In Genesis 38, we read the account of Judah and his daughter-in-law, Tamar. For reasons we won't go into here (read the biblical account if you want the sordid backstory), Tamar, who'd been widowed and was desperate for a child, lured Judah into a tryst by posing as a cult prostitute. It's possible she fooled Judah by pretending to be a *harimtu*, one of the devotees of Inanna, probably known as Astarte by Judah's time. Nine months after their roadside encounter, Tamar gave birth to twins, Perez and Zerah. Among the descendants of Perez were David, king of Israel, and, almost 1,700 years later, one Jesus of Nazareth.

The point is that Judah founded the tribe that produced the Messiah by sleeping with his widowed daughter-in-law because he thought she was prostitute serving a pagan goddess. If Jacob and his sons had remained in Canaan any longer than they did, how long would it have been before they degenerated into just another bunch of sexually immoral pagans?

In Egypt, they weren't likely to be pulled into the polytheistic culture of the host country. When Joseph hosted a meal for his brothers (before he revealed his true identity):

> They served him by himself, and them by themselves, and the Egyptians who ate with him by themselves, because the Egyptians could not eat with the Hebrews, for that is an abomination to the Egyptians. (Genesis 44:32, ESV)

The Egyptians were, well, racists. Because of their belief in their superiority, they wouldn't associate with Hebrews except when necessary. Egyptian texts from that period of history are filled with derogatory references to "sand dwellers" and "Asiatics."

But the racial discrimination of the Egyptians served God's purposes. For 215 years, the Israelites were isolated from outside cultural influences while they were busy being fruitful and multiplying. More important, they were cut off from the temptations of foreign gods. In Egypt, over the four generations between Jacob and Moses, Israel became a distinct people.

The time had finally come. God called Moses back to Egypt to bring His people, Israel, to the place He'd claimed as His own—Canaan.

Yahweh chose to convince Pharaoh and the Egyptians to not only let Israel leave, but to encourage them to go. He did it by hardening Pharaoh's heart through a series of increasingly severe trials until the people of Egypt must have been *begging* Pharaoh to let His people go.

There are several studies you can find online that draw links between the ten plagues Yahweh inflicted on Egypt and specific Egyptian gods. For example, the first plague turned the Nile River to blood. This is said to have been directed at Hapi, the god of the annual Nile flood. Plague number two, frogs, was aimed at Heqet, a fertility goddess worshipped since the early dynastic period—the time of Narmer and the first kings of Egypt, about 1,500 or 1,600 years before Moses.

Those match up well enough, but when we get to the third and fourth plagues, the connections are iffy at best. The plague of lice or gnats, depending on the translation you read, doesn't match up well with any known Egyptian god. The plague of flies is paired by some with Khepri, a god of creation. But Khepri had a scarab beetle for a head, so that's not a good match, either.

Some of the pastors and teachers who've published these studies are very intelligent people whom I respect. However, and with all due respect to those pastor-teachers, they've overlooked an even bigger supernatural conflict. Understanding that confrontation will show you why trying to link the ten plagues to specific Egyptian gods is looking in the wrong direction. More accurately, it's looking at entirely the wrong pantheon.

Yes, Yahweh demonstrated with the ten plagues that His power was superior to that of the gods the Egyptians trusted to keep the Nile flowing and the crops growing. And we know for a fact that Yahweh put a hurt on the gods of Egypt the night He took the lives of Egypt's firstborn.

How do we know? He told Moses.

> For I will pass through the land of Egypt that night, and I will strike all the firstborn in the land of Egypt, both man and beast; **and on all the gods of Egypt I will execute judgments**: I am the LORD. (Exodus 12:12, ESV, emphasis added)

How likely is it that Yahweh told Moses that He was about to punish imaginary beings represented by idols of wood and stone? What would be the point? How would that establish His power and glory?

No, something happened in the spirit realm on the night of the Passover. When Yahweh passed through the land of Egypt, taking the lives of firstborn humans and animals, He simultaneously carried out His sentence on the bene elohim, the entities who had rebelled and made themselves gods in Egypt.

Here's a fascinating detail we never hear about in church: It appears there was a very old tradition in Egypt, an ancient myth dating back centuries before the Exodus, that a day was coming when the first-born of

Egypt would die. The pyramids of the fifth Dynasty king Unas, c. 2350 B.C., and the sixth Dynasty king Teti, c. 2320 B.C., are inscribed with this line from a well-known inscription called the "Cannibal Hymn":

It is the king who will be judged with Him-whose-name-is-hidden on this day of the slaying of the first-born.[15]

Similar phrases are found on other coffins from Egypt's Middle Kingdom, including a variant that reads "this **night** of the slaying of the first-born." Some scholars believe the context of the Coffin Texts and the Cannibal Hymn points to the first-born belonging to the gods, although that's not a view shared by all Egyptologists.

What does it mean? Scholars aren't sure. But it seems that by the time of the Exodus, there was a very old tradition in Egypt of a future nightmare event when the firstborn would be killed.

Consider this possibility: Maybe the Coffin Texts and the "Cannibal Hymn" were an ancient warning to Egypt of that coming day of judgment. And forty years earlier on Mount Sinai, Yahweh revealed to Moses that *He* was Him-whose-name-is-hidden, I AM WHO I AM—the One who would someday fulfill the prophecy of the slaying of the firstborn.

That's speculation, of course, but fascinating. And we're not at the best part yet.

You know the outline of the story. Pharaoh finally relented and let the Hebrews go. Moses and the Israelites threw their things together and left in a hurry. Led by the Angel of Yahweh (who was also Yahweh), the Israelites fled, heading east out of Egypt.

Their route didn't take the most direct path toward Canaan. Yahweh sent them south and east "lest the people change their minds when they see war and return to Egypt." It's likely that sending them toward the land of the Philistines would have brought them past Egypt's border defenses, forts, or military posts guarding the major routes into the kingdom.

Scholars today, 3,500 years later, still argue about where the Red Sea crossing occurred. We won't get into it here. If it hasn't been settled by now, we're not going to put the question to bed in a couple of paragraphs. Besides, that's not important right now. What matters is what Yahweh told Moses to do next.

> Then the LORD said to Moses,
> "Tell the people of Israel to turn back and encamp in front of Pi-hahiroth, between Migdol and the sea, **in front of Baal-zephon; you shall encamp facing it**, by the sea." (Exodus 14:1–2, ESV, emphasis added)

Okay, this begs some questions: Why did God tell Moses to turn back? Why did He command Moses to camp facing Baal-zephon? What *is* Baal-zephon? And mostly, *what was Ba`al doing in Egypt?*

You know that Ba`al was the Canaanite storm-god and the king of the Canaanites' pantheon. He's mentioned in the Bible from the book of Exodus through the Gospels. Ba`al, which is properly pronounced *bah-awl* with a glottal stop like, "Uh-oh," was the main thorn in the side of the Israelites, especially those who were faithful to Yahweh, for the next 1,500 years, all the way down to the time of Jesus.

But during the Second Intermediate Period in Egypt, roughly 1750 to 1550 B.C. (give or take a hundred years), foreigners from Canaan called the Hyksos ruled northern Egypt. Their capital was at a city called Avaris in the Nile delta, and they worshipped the gods of the Canaanite pantheon, headed by Ba`al.

That's why the well-intentioned efforts to identify Egyptian gods as targets for the plagues are looking at the wrong pantheon. During the period of history when the Israelites were in Egypt, from about 1665 B.C. to 1450 B.C., the country was divided. Native Egyptian rulers only controlled the southern part of the country, which, oddly enough (to us Americans), was called Upper Egypt. Their capital was at Thebes. Lower (northern) Egypt was under Hyksos control, and their gods were the ones worshipped by the Semitic inhabitants of the Levant and Mesopotamia.

The Hyksos might have been Semitic, although some scholars think they were Indo-European invaders who came down from what is now Turkey and adopted Semitic names. That's not unheard of. The German House of Saxe-Coburg and Gotha changed its name to Windsor, which sounded more appropriate for the royal family of England—especially in 1917, when King George V changed it at the height of World War I. Similarly, the Hyksos nobility may have adopted Semitic names when they moved into Canaan from the north to connect with their subjects.

So—did you know there was once an Egyptian king named Jacob?

No, probably not *that* Jacob, but it's true. Yaqub-har ruled in northern Egypt sometime in the seventeenth or sixteenth century B.C. during the Second Intermediate Period. The suffix *har* means "mountain" in Canaanite and Hebrew, so whoever he was, about 3,600 years ago, part of Egypt was ruled by a king named Jacob of the Mountain. Some scholars have suggested that a proper reconstruction of the king's name in Semitic would be Yaqub-haddu. Haddu was another form of the name of the West Semitic storm god Hadad, who we know by his later title, Lord—Ba`al. (And a few scholars translate this king's name Yaqub-baal or Yaqub-el).

This style of name is called "theophoric" because it includes the name of a god. It was common in most of the ancient Near East beginning in the third millennium B.C., and it continued well into the biblical era. In fact, theophoric names help scholars tell Amorite names, which often incorporated Dagan (Dagon), Līm, or Haddu, from, say, Hebrew names, which were often built around YHWH, as in Hezekiah (Ḥizkiyyahu— "strengthened by Yahweh"), or El, as in Michael, Samuel, Daniel, etc.

Interestingly, another scarab bearing the name Yaqub-har in Egyptian hieroglyphic was found near Haifa, Israel. This piece was dated to about 1730 B.C., possibly seventy or eighty years too early to be the Hyksos king. The most likely explanation, scholars think, is that the scarab belonged to another Jacob who lived about a century before the Hyksos came to power, probably a local Semitic prince in Canaan under the overlordship of Egypt—and that Jacob *might* have been Jacob the patriarch. We'll probably never know for sure.

We also uncertain about whether the Hyksos were still around at the time of the Exodus. At some point, possibly while the Israelites were still living in Goshen in the Nile Delta, an Egyptian king decided he'd had enough and brought an army north to drive the Hyksos out of the land. The revolution began under Seqenenre Tao, whose badly battered mummy shows that he probably lost his last battle against the Hyksos. His sons, Kamose and Ahmose, carried on the fight, with Ahmose finally driving out the hated Asiatics after a war that appears to have lasted for at least twenty years.

You've noticed the similarity between the names of Seqenenre's sons and Moses. Yes, Moses is an Egyptian name.

It appears the Hyksos left Avaris under a negotiated truce, but Ahmose apparently changed his mind about the deal. He chased the Hyksos toward Canaan, catching up with them at a Hyksos stronghold called Sharuhen, a town either in the Negev Desert, south of Israel, or near Gaza (Joshua 19:6 puts it in the territory of the tribe of Simeon). After a three-year siege, Ahmose took the town and razed it, ending Canaanite political influence in Egypt once and for all.

The revolution was apparently provoked by an insulting letter sent by the Hyksos king Apepi (also called by the Greek form of his name, Apophis) to Seqenenre complaining about the snoring of the hippos in the sacred pool at Thebes. Now, since Avaris was about 375 miles from Thebes, it's not like Apepi was really losing sleep because of the sacred hippos, but it appears Seqenenre couldn't ignore the insult.

The animosity between the Hyksos and the Egyptians is hinted at in the names of the kings. Seqenenre means "who strikes like Re," meaning Ra, the sun god. Both Kamose and Ahmose adopted prenomen (one of the "great five names" of Egyptian rulers) that included the name of Re/Ra.

Apepi ("man of Apep"), while he also had the name of Re in his prenomen, because that was the formula back in the day, was named for Apep/Apophis, the evil supernatural serpent that tried to eat the solar boat of Ra every night. Set, during this period of Egyptian history, was still a hero; he rode the prow of Ra's boat and defended it against Apep when it went below the horizon, a sort of cosmic Captain Ahab doing battle night after night with a divine Moby Dick to ensure the next day's dawn.

Maybe the Hyksos merged Set into Ba`al, the chief god of their pantheon, as a claim of superiority—a message to the Ra-worshiping Theban kings that *our* god keeps *your* god from being snake food, a claim reinforced by the Hyksos king adopting the name of Ra's mortal enemy.

Quick historical detour: The Egyptian pharaoh Akhenaten, formerly Amenhotep IV, is often credited by scholars with inventing monotheism when he established the sun god, Ra, as the only god to be worshiped. Some then go so far as to suggest that Moses stole the idea from Akhenaten— which, of course, is another supernatural PSYOP.

However, scholars have found that the Hyksos king Apepi was a monotheist two hundred years before Akhenaten. A literary text from the nineteenth dynasty, the dynasty of Seti and Ramesses, records that Apepi decreed the sole worship of Ba`al-Set in Lower Egypt.

> It once happened that the land of Egypt was in misery, for there was no lord [as] (sole) king. A day came to pass when king Seqenenre' acted as a Ruler of the Southern city. Pestilence was in the town of the Asiatics, while (the) great one Apophis was in Avaris, as the entire land paid him tribute, delivering their taxes— also the north—bringing every sort of good product of the Delta. Then, **king Apophis adopted for himself Seth as lord, and he refused to serve any god that was in the entire land except Seth.**[16] (Emphasis added)

In other words, it's possible the Egyptian war for independence wasn't triggered by Apophis' insult of the sacred hippos of Thebes, but because Apophis engaged in what native Egyptians considered rank heresy—the worship of one, and only one, god. (The successors of the monotheist Akhenaten certainly did everything they could to erase *his* name from history.)

As best we can tell, it appears the principalities and powers who influenced the Hyksos brought them to Egypt at precisely the time of the Israelites' sojourn, and everything else that happened, including the war for Egyptian independence, was part of a bigger plan: Genocide, something the house of Israel has survived many times now.

Consider: Just as Joseph was kidnapped by traders and carried off to Egypt, laying the groundwork for his family to follow and grow into a nation, a group of Jacob's Semitic-speaking, Ba`al-worshiping neighbors *also* arrived in Egypt and took over the government from the natives.

Coincidence?

No. Analyzing history through a biblical lens, the Second Intermediate Period in Egypt was apparently a plan by the Fallen to bring a specific group to power in the land where Yahweh had led the Israelites. The purpose was to destroy the people that God had chosen for Himself.

Regardless of how it started, the long war between the native Egyptians in the south and the foreign Hyksos in the north could explain why the Israelites fell out of favor. If power changed hands from Semitic overlords to native Egyptian kings, the Semitic-speaking Hebrews would have been seen as potential enemy collaborators. The narrative in Exodus fits the political situation of the mid-fifteenth century B.C.

> Now there arose a new king over Egypt, who did not know Joseph. And he said to his people, "Behold, the people of Israel are too many and too mighty for us. Come, let us deal shrewdly with them, lest they multiply, and, if war breaks out, they join our enemies and fight against us and escape from the land." (Exodus 1:8–10, ESV)

The dates are uncertain and scholars disagree over which Egyptian king was the pharaoh of the Exodus, although it's pretty much agreed that it wasn't Ramesses II, better known as Ramesses the Great—or Yul Brynner, if you're old enough to remember the movie. Ramesses lived and ruled about two hundred years later, between 1279 B.C. and 1213 B.C.

It may seem counterproductive for Ba`al to lead the Hyksos into

Egypt, set them up as kings, and then allow them to be run off again by native Egyptians within a hundred years or so. But the Egyptians' hatred of the Israelites was a byproduct of the Hyksos intervention. Besides, despite what you've seen in movies over the years, the worship of Ba`al and other Semitic gods continued in northern Egypt long after the Hyksos were driven out.

During the reign of the Hyksos, Ba`al was identified with the god Set, the Egyptian god of storms, chaos, the desert, and foreigners—a good fit with Ba`al, apparently. The Hyksos adopted Set and blended the two into a single entity.

Today, we think of Set, sometimes called Seth or Sutekh, as an evil god who cut his brother Osiris into fourteen pieces. He was usually depicted as a bizarre character with a human body and an anteater-like head. But as we've mentioned, Set wasn't always a villain. In the time of the Hyksos and for a few centuries after, Set helped the sun god, Ra, by defeating the evil serpent Apep/Apophis, the embodiment of chaos, which tried to eat Ra's solar boat every night as it disappeared over the horizon. In this tale, Set's nightly victory over Apep echoed the Semitic myth in which Ba`al vanquished the chaos god of the sea, Yam, and his sea-dragon servant, Lotan.

Points to you if you noticed the similarity between the name Lotan and the Bible's Leviathan. It's a classic PSYOP by the Fallen, claiming the deeds of Yahweh as their own.

When Apep delayed Ra's boat, there were storms; when he ate it, there was an eclipse. But Apep's victories were always temporary, and every evening Set was back on the front of the boat to spear the serpent.

Later in Egyptian history, though, after being conquered by the Nubians, Assyrians, and Persians one after another between 800 B.C. and 525 B.C., the god of foreigners was no longer welcome around the pyramids. That's why later Egyptians considered Set evil.

The worship of Ba`al-Set continued in northern Egypt for at least four hundred years, long after the Hyksos were run out of the country. In fact, the pharaohs of the Ramesside dynasty, who were the first Egyptian

kings to be called pharaoh (which means the term wasn't used in the days
of Joseph or Moses), were worshipers of Ba`al-Set. The father of Ramesses
the Great was named Seti I—literally, "man of Set." Several other kings
of the Ramesside period, including Seti II and Setnakht ("Set is strong"),
were also named for the god.

There is some speculation that because Ramesses the Great was a
redhead (true!), his family may have been descendants of the Hyksos
invaders. Whatever the reason, Ramesses II set up a stela at Pi-ramesses,
near the site of the old Hyksos capital Avaris, to commemorate the four
hundredth anniversary of Set doing…something. Scholars aren't sure of
what, exactly. The unimaginatively named Year 400 Stela portrays Seti I
presenting wine to Set, who's depicted like the images of Ba`al found in
the Levant, with a human head instead of the more familiar anteater-head
from later Egyptian history.

The Year 400 Stela features Seti I (right), father of Ramesses
the Great, and the god Set (left), who was identified with
Ba`al in Egypt for most of the second millennium B.C.

Since the stela points back four hundred years to about 1650 B.C., it could mark the arrival of Ba`al-Set in the Nile Delta. Counting back 215 years from the Exodus date of 1450 B.C., this is just about the time Jacob and the family arrived in Egypt.

Again, we ask: Coincidence?

When Yahweh led the Israelites out of Egypt, He ordered them to turn around and camp on the shore of the Red Sea facing something called Baal-zephon all night. Why? Specifically so they'd cross the Red Sea right in front of it.

Here's the funny part: As worshipers of the desert-god Set, the Egyptian army probably thought they had the Israelites *right where they wanted them.*

> For Pharaoh will say of the people of Israel, "They are wandering
> in the land; the wilderness has shut them in." (Exodus 14:3, ESV)

Caught in front of Baal-zephon between areas controlled by their god, Ba`al-Set, master of the sea and god of the desert—the Egyptians must have figured the Israelites were hopelessly trapped!

Which brings us back to our earlier question: What was Baal-zephon, and why was it so important? Why did God tell Moses to turn around so this confrontation happened right there?

Here's why: The name of Ba`al's holy mountain, which is more than five hundred miles north of the Nile delta, was Mount *Zaphon.*

Hmm. Zephon, Zaphon. Same name, different transliteration into English. Coincidence?

No! The Red Sea crossing was a supernatural *smackdown!* Ba`al was the god of storms, the god who vanquished the primordial chaos god of the sea, Yam. Because of this, Ba`al was the god of maritime navigation and the patron god of sailors.

So Yahweh didn't just deliver the Israelites "out of the hand of pharaoh," *He delivered them out of the hand of Ba`al.* And to make sure nobody misunderstood the message, He did it in front of a site dedicated to Ba`al, and by mastering the sea—Ba`al's domain.

This was a called shot! Just like Babe Ruth at Wrigley Field in the 1932 World Series, the confrontation at the Red Sea was engineered by Yahweh to serve notice to the Fallen: *My people are freed from their bondage and now we are coming for you.*

If there had been a microphone to drop in heaven, that's when it would have hit the floor. *Boom.*

That was the reason for the crossing. Yahweh used it to demonstrate His power, yes, but for a specific purpose—to make it crystal clear to the Israelites that I AM was unparalleled, unchallenged, and sovereign. It was a demonstration of His authority over the divine entities who'd chosen to abuse the responsibilities He'd given them after Babel. And it was a clear message to the gods that the days of their rebellion were numbered.

But because we Christians have desupernaturalized the Bible, we never hear the Red Sea crossing explained in these terms.

One more thing: To this day, no one is sure exactly where the Red Sea crossing took place. Since the Hebrew words translated Red Sea, *yam suph*, actually mean "sea of reeds," some believe the crossing took place at one of the marshy areas in the Nile delta. That makes it easier to believe the miracle, apparently. They point out that a strong east wind blowing all night could maybe, possibly, cause the waters of a marsh to recede enough to allow people to cross—as if the Creator who spoke the universe into existence couldn't whip up an east wind strong enough to part the Red Sea.

Setting aside that disbelief: Can we just stop for a minute and think about how incredibly silly it would have been to put a site sacred to the patron god of sailors at the edge of a marsh?

The Bible says what it means. Yahweh rescued His people at the Red Sea. And it was there—it *had* to be there—because it was a message to Ba`al and the rest of the spirit realm.

———

So the Israelites escaped Egypt and made their way across the desert to Sinai, the next important holy mountain on our list. Sinai is where Moses would receive the Law. Yahweh had called Israel to His holy mountain, and He was about to open the next phase in the long war against the Fallen.

It's important to note the instructions and warnings God gave to Moses when the people arrived at Sinai:

> The Lord said to Moses, "Go to the people and consecrate them today and tomorrow, and let them wash their garments and be ready for the third day. For on the third day the LORD will come down on Mount Sinai in the sight of all the people. And you shall set limits for the people all around, saying, 'Take care not to go up into the mountain or touch the edge of it. Whoever touches the mountain shall be put to death.'" (Exodus 19:10–12, ESV)

To put it bluntly, Mount Sinai was sacred to the point of being taboo. This was extreme holiness, the kind of holy that Mount Hermon only pretended to be. Sinai was, at that moment in time, the holy mountain of God, the mount of assembly. And did you notice that Yahweh came down to the mountain on the third day?

Coincidence?

Now, just imagine the scene from the perspective of the people gathered at the foot of the mountain:

> On the morning of the third day there were thunders and lightnings and a thick cloud on the mountain and a very loud trumpet blast, so that all the people in the camp trembled. Then Moses brought the people out of the camp to meet God, and they took their stand at the foot of the mountain. Now Mount Sinai was wrapped in smoke because the LORD had descended on it in fire. The smoke of it went up like the smoke of a kiln, and the whole mountain trembled greatly. And as the sound of the trumpet grew louder and louder, Moses spoke, and God answered him in thunder. (Exodus 19:16–19, ESV)

The people trembled? No doubt! This was straight out of a science-fiction film, the moment the alien mothership breaks into earth's atmosphere! (Of course, there are ancient astronaut believers who argue that's exactly what the people saw. Nonsense.)

Well, Moses got the Law, and then Yahweh told him to come back up the mountain to receive instructions for building the tabernacle. But first, Yahweh did something remarkable.

> Then he said to Moses, "Come up to the LORD, you and Aaron, Nadab, and Abihu, **and seventy of the elders of Israel**, and worship from afar. Moses alone shall come near to the LORD, but the others shall not come near, and the people shall not come up with him."…
>
> **Then Moses and Aaron, Nadab, and Abihu, and seventy of the elders of Israel went up, and they saw the God of Israel.** There was under his feet as it were a pavement of sapphire stone, like the very heaven for clearness. And he did not lay his hand on the chief men of the people of Israel; **they beheld God, and ate and drank.** (Exodus 24:1–2, 9–11, ESV, emphasis added)

Do you get the significance? For the first time since Eden (not including the burning bush incident, of course), humans were in the presence of Yahweh on His holy mountain. This was an early taste of the reconstituted divine council.

Here's the other key point: How many elders of Israel were there? How many nations did God create after the Tower of Babel incident? How many sons of El in the assembly on Mount Hermon?

Seventy, seventy, and seventy.

Coincidence?

No! It was another message to the Fallen: *A day is coming when my people will again take their place in the divine council.*

There are parallels between Eden, Mount Sinai, and the tabernacle (and later, the Temple on Mount Zion) rich enough in meaning to deserve entire books—and indeed, entire books have been written on the topic. We'll briefly list a few just to make the point, again, that there are no coincidences with God.

The main takeaway is this: The tabernacle and the Temple were representations of Yahweh's holy mountain. As with Eden, the entrances to the tabernacle and Temple were on the east side. As with Sinai, the manmade structures were divided into three zones of increasingly restricted access: The base of the mountain was like the outer court with an altar beyond which the Israelites could not pass. Midway up the mountain represented the Holy Place: Moses, Aaron and his sons (the priests), and the elders of Israel climbed halfway up Sinai to eat a meal with Yahweh; likewise, only the priests could enter the Holy Place in the tabernacle and Temple. And the summit of Sinai was akin to the Holy of Holies, where the visible manifestation of Yahweh descended from heaven. It was open on Sinai only to Moses, and in the tabernacle and Temple only to the high priest.

Ba`al wasn't finished, of course. His motto, apparently, was if you can't beat 'em, lie to 'em. Moses was on the mountain for forty days. God gave him a lot of information—the Law and specifications for the tabernacle, including the Ark of the Testimony. The people got antsy during Moses' long absence, no doubt with some encouragement from Ba`al and his minions. Under pressure from the Israelites, Aaron made a golden calf to represent the gods who brought them out of Egypt. And his explanation was something that could have come from a television sitcom writer:

> And Aaron said, "Let not the anger of my lord burn hot. You know the people, that they are set on evil. For they said to me, 'Make us gods who shall go before us. As for this Moses, the man who brought us up out of the land of Egypt, we do not know what has become of him.' So I said to them, 'Let any who have

gold take it off.' So they gave it to me, and I threw it into the fire, and out came this calf." (Exodus 32:22–24, ESV)

"I threw the gold in the fire and out came this calf." And Aaron was supposed to be the brother with the gift of public speaking!

You probably see this coming, but guess which gods in the Canaanite pantheon were typically represented by calves or bulls?

Ba`al and El. Coincidence?

Not on your life.

SUMMING UP

Abram was called by Yahweh from the northern Mesopotamian city of Ura and led to Canaan. While there, Yahweh made a covenant with the aging and childless Abram, promising that his descendants would inherit the land from the border of Egypt to the Euphrates. God also told Abram that his descendants would be afflicted for four hundred years, but would return from a sojourn in a foreign land in the fourth generation because the iniquity of the Amorites was not yet complete.

Moses was likewise called by Yahweh and tasked with bringing His people out of Egypt, where they had grown into a generally cohesive and distinct nation. The ten plagues of Egypt culminated with the slaying of the firstborn, a night on which Yahweh executed judgments on all the gods of Egypt. Then he arranged a confrontation on the shore of the Red Sea, specifically to demonstrate His power over the storm-god and vanquisher of the sea, Ba`al.

And to ensure that the bene elohim received the message, Yahweh invited Moses, Aaron, and the seventy elders of Israel up to His holy mountain, Sinai, where they ate a meal with God and saw Him face to face—a specific message to the Fallen that humans would one day retake their original place in the divine council.

One more odd bit of information to consider: As we discussed, Ba`al-Set was the chief god of Lower Egypt while the Israelites were there. While Set was equated with Ba`al, he was also identified with the Greek god of chaos, Typhon. Typhon was the deadliest, most dangerous entity outside of Olympus.

Frankly, this is confusing. In Greek myth, Typhon fought an epic battle with Zeus and almost won. (Part of the battle, interestingly, was fought at Mount Zaphon, called Mount Kasios by the Greeks.) Zeus, as we'll show you in the next chapter, was Ba`al.

Weird, right? Zeus = Ba`al = Set = Typhon…who nearly killed Zeus. It doesn't make sense, but the Fallen don't care what you believe as long as it isn't true.

The point: Researcher and author Peter Levenda, in his book, *The Dark Lord*, builds a good case that occultist Aleister Crowley, founder of the magickal system called Thelema, and horror fiction author H. P. Lovecraft, who created the Cthulhu mythos, drew their inspiration from the same supernatural source. Levenda identifies that source as Set-Typhon, the dark lord of chaos.

A recurring theme in Lovecraft's stories is the complete irrelevance of mankind in the face of cosmic horrors that lurk in the universe. To Lovecraft, creation was occupied by the Great Old Ones, a pantheon of deities who once ruled the earth but who have since fallen into a deathlike sleep. The physical forms of these horrible beings were so monstrous that the human mind was incapable of processing them. Just *seeing* them was enough to drive people insane. Cheery, eh?

After Crowley's death, one of his acolytes, Kenneth Grant, claimed he'd discovered an extraterrestrial Sirius/Set energy current. Grant adapted Crowley's system of ritual magick into a new one he called Typhonian.

So the octopus-headed monster Cthulhu may have been H. P. Lovecraft's nightmare vision of the chaos-god Set, otherwise known as Ba`al.

And *that* is the god who arrived in Egypt just in time to enslave the Israelites.

Coincidence? No.

So let's dig up some more intel on this god.

5

ZAPHON

As we mentioned in the previous chapter, West Semitic cosmology placed Ba`al's palace on Mount Zaphon. It's an imposing peak on the seacoast just north of the border between Syria and Turkey, rising about 5,800 feet above the Mediterranean Sea. Today it's called Jebel al-Aqra, but in the classical world everybody knew it was Ba`al's mount of assembly.

Ba`al was the main enemy of Yahweh in the Old Testament. One could argue that he was even more of a villain in the Bible than Satan, who's only mentioned in fourteen Old Testament verses, in the books of 1 Chronicles, Job, and Zechariah. Ba`al, on the other hand, appears 106 times in 88 verses (including personal and place names, like Baal-zephon). He arrives in Exodus and he's still around in the New Testament. Jesus discussed him with the Pharisees (see Matthew 12:22–28).

The Bible never explains that Ba`al isn't a name, it's a title. It means "lord," in the same way we Christians usually say Lord (and our Bibles read Lord) instead of saying Yahweh or spelling out YHWH.

Stela of Ba`al wielding his mace
and thunderbolt found at Ugarit,
now on display at the Louvre.

That's rather arrogant, when you think about it. And Ba`al wasn't the only one; over time, Babylon's chief god Marduk came to be known as Bel, which also means "lord."

Ba`al has had a long career. He's not recognized as an active deity in our world today, to the best of our knowledge. There may be a few here and there who try to recreate the old religion of the ancient Canaanites, but like aging pop divas, the old gods reinvented themselves as time and people moved on. We've already seen how Sumerian Inanna became Babylon's Ishtar, the Semitic Astarte, Aphrodite of Greece, and Venus of Rome.

It's possible, too, that some of those historic geopolitical changes—such as Rome replacing Greece, which replaced Persia, which replaced Babylon, which replaced Assyria, etc.—resulted from power struggles between the gods themselves. There is nothing in Scripture that tells us, but there is no reason to think the seventy *bene elohim* weren't all as ambitious as the divine rebel condemned by Isaiah, Helel ben Shachar.

And there are reasons to believe that the Fallen aren't above using humans as cannon fodder to advance their agenda. It's a win-win for them. They get a blood sacrifice—the blood is the life—and spread hatred among men like a virus, which is exactly the opposite of our Christian command to love our enemies.

Over time, Ba`al replaced the original name of the West Semitic storm-god Hadad (or Haddu), who was also called Adad/Addu by the Akkadians. Ba`al has also been identified with the Hurrian god Teššub, the Hittite god Tarhunt (or Tarhun), the Urartian god Teisheba, the Sumerian god Iškur, the Amorite god Amurru (incorrectly, as it happens), and others. As you know from the previous chapter, you've already met Ba`al as Zeus and Jupiter.

The earliest of these storm-god manifestations was Iškur (the š sounds like the "sh"). Iškur makes a cameo appearance in *Enmerkar and the Lord of Aratta*, bringing rain to the land of Aratta just in time to break a famine, which gave its ruler the courage to reject Enmerkar's demand for building materials.

However, Iškur was a minor deity in the Sumerian pantheon. That's logical; in a land that's mostly desert, a god of storms and life-bringing rain wasn't as important as Enki, who controlled the fresh waters of the Tigris and Euphrates. Irrigation was far more important than rainfall for growing crops in southern Mesopotamia. Farther north, where rain was key to agriculture, Iškur played a larger role in the pantheon.

The chief god Enlil and his son, the war god Ninurta, also had storm-god characteristics. This left Iškur looking like a pale imitation of the others, a third-tier superhero in a comic book universe. He was believed to be the son of Anu, although some traditions claimed that Enlil was his father. This could mean that Iškur wasn't a native Sumerian deity but an import from Sumer's Semitic-speaking neighbors. As the patron god of Karkara, a minor city near Uruk in southern Mesopotamia, Iškur's cult center didn't have enough political clout to elevate him into the upper rank of the pantheon, as Babylon later did for Marduk.

The etymology of Iškur's name is uncertain. That could mean it's archaic Sumerian, an obsolete word whose meaning has been lost to the ages, or it might have been borrowed from a language that's neither Sumerian or Semitic. We can only speculate. While the Sumerian word *kur* means "mountain" (and "dragon"), which is intriguing considering the importance of holy mountains in history, we should be careful not to draw conclusions just because part of a word looks or sounds like a word that's possibly unrelated. That's like picking your favorite definition from a list of possible choices in *Strong's Concordance* to make a Bible verse say what you want it to say.

However, this does indicate that the worship of Iškur extends back into prehistory. The first written evidence of the cult of Iškur comes from the cities Lagash and Adab in the middle of the third millennium B.C. Adab was near Karkara, site of Iškur's temple, the House of the Big Storms. Unfortunately, the ruins of Karkara haven't been discovered yet, so we don't know much about his cult or the local traditions about him.

Iškur was typically shown standing on the back of, or in a chariot pulled by, a lion-dragon. He's often shown with an axe in one upraised hand and lightning bolts in the other. Based on prayers and rituals that have been preserved, it appears Iškur, like most of the Mesopotamian gods and goddesses, could be good or bad depending on his mood. The storm-god brought necessary rain for the crops needed to survive, but he could also destroy fields with wind, hail, and flood.

He first appeared under the name Hadda at Ebla around 2500 B.C., and as Adad in Old Akkadian texts, the period of history that began with the empire of Sargon the Great around 2330 B.C. We know they're the same because the Sumerian logogram for Hadda/Adad was the same used for Iškur (^{d}IM). As god of the heavenly waters, Iškur was sometimes praised as the twin brother of Enki. This may be an early example of the "twin gods" motif in ancient mythology. Sometimes they're brother-sister twins, such as Apollo and Artemis; sometimes divine-mortal twins, like

Castor and Pollux; and sometimes, as in this case, a pair of brothers, one more important than the other—at least in the early days. At that point in history, it would not have been a compliment for Enki, the god responsible not only for the sweet waters of the *abzu* but for the fundamental laws of civilization, the *mes*, to be compared with a minor god like Iškur.

Now, we can only guess at the true nature of these spirit beings. Our ability to see into that realm is limited and the spirits who occupy it don't always tell the truth. All the myths and legends about the pagan gods, from Sumer to the present day, must be taken with a grain of salt. The stories are meant to deceive, confuse, and divide us—in other words, they're PSYOPs. The best we can do is see the outlines of the big picture by looking at long trends and historic outcomes—the fruit of the behind-the-scenes activity of the supernatural players.

Iškur-Adad, and later Ba'al, played a key role in the cosmology of Mesopotamia for the next 2,500 years. His importance to the cultures of the region grew with the political influence of the Amorites, who may have introduced Iškur into Sumer, reaching its peak in the second millennium B.C. and lasting through the time of Jesus.

At Mari, which was located on the Euphrates about seventy-five miles southeast of the modern city of Deir ez-Zor, a powerful but short-lived Amorite kingdom grew from the ashes of an older state that had been smashed by Sargon the Great. The name Haddu often appears in texts found at Mari alongside Dagan, a grain god who was the chief deity of the middle Euphrates region, and Itūr-Mēr as one of the three great gods of the kingdom of Mari.

Quick detour: There is a widespread but mistaken belief that Dagan, known to Bible readers as Dagon, god of the Philistine city of Ashdod, was a fish-god whose priests wore fish-head hats that inspired the mitre worn by Roman Catholic popes. This is not the case. Dagan's false reputation as a fish-god is largely due to Alexander Hislop's poorly researched 1858 book, *The Two Babylons*. The relevant image in the book, figure 48, is an image we included earlier in this book, a cylinder seal impression showing two apkallu and a human king standing before the Tree of Life and a winged solar disk. One of the apkallu, the fish-man type, was incorrectly identified by Hislop as Dagon. Why, we'll never know.

But I digress.

As political control of Mesopotamia shifted from Akkad to Ur to Babylon between 2150 B.C. and 1900 B.C., powerful Amorite kingdoms in Assyria, Mari, and Yamḥad (modern Aleppo, Syria) emerged alongside the old Babylonian empire. At the same time, and not coincidentally, the storm-god's identity gradually shifted from Sumerian Iškur to Akkadian Adad while his importance in the pantheon grew. By the time Hammurabi brought Babylon to the greatest extent of its power, Adad was firmly established among the great gods of Mesopotamia.

Meanwhile, the bull replaced the lion-dragon as the animal associated with Adad, a symbol more familiar to those who know Ba`al from the Bible.

Adad was also known during this period as a god of law and justice. Oaths were often sworn before a symbol of the god, and Adad was often called upon with the sun-god Šamaš (Shamash) as witnesses in legal matters. Adad and Šamaš were also considered the gods of divination; specifically, extispicy—revealing the future through the patterns on animal entrails, especially the livers of sacrificed sheep and poultry. Why Adad and Šamaš were the experts, scholars don't know. One can imagine the rebellious elohim having a good laugh over selling humans on *that* idea.

Another quick rabbit trail: Did you ever wonder why God specifically told Moses multiple times to burn "the long lobe of the liver" of sacrificial animals?[17] This is why. Yahweh was making it clear there was to be no divination among His people.

While texts from the Canaanite city-state Ugarit name Mount Zaphon as the site of Ba`al's palace, the capital of Yamḥad, Halab (Aleppo), was known across the ancient Near East as the City of Hadad. The storm-god's sanctuary there, which lies beneath a massive citadel in Aleppo's old quarter, dates to the Early Bronze Age, the mid-third millennium B.C. (~2500 B.C.), and it was in use until about the ninth century B.C.

The importance of the storm-god of Aleppo is highlighted by the evidence of his cult across Mesopotamia and the Levant, from Nuzi, east of the Tigris River, to Hattuša, capital of the Hittite empire, in what is now north-central Turkey. To be fair, though, the Hittites worshipped just about everybody in the spirit realm. They made a practice of adopting the gods and goddesses of all the lands they conquered to the point that their pantheon became unwieldy with foreign gods. It appears that the Hittites tried to maintain their empire by convincing the gods that it was in their best interests to "remain in the Hatti land", home of the best and purest sacrifices, festivals, and temples.

Still, there is no question that by the old Babylonian period (1700s B.C.), Hadad was, other than Marduk, the preeminent god in Mesopotamia. Emissaries from Elam, today's northwest Iran, traveled to Halab to present a bow as a gift to Hadad.

Yamḥad certainly enjoyed the benefits of the storm-god's presence in its midst. The king was called "beloved of Hadad" and the kingdom was the Land of Hadad. The god bestowed kingship and assigned territory to kings, even to those outside the borders of Yamḥad. Before the Amorites swept into southern Mesopotamia and took control from the native Sumerians in the early second millennium B.C., that power was restricted to Enlil. (After the rise of Babylon, various deities besides Hadad, including Marduk, Dagan, and the moon-god Sîn, claimed king-making authority at different times and places.)

If you're a reader of the Bible, you recognize that this is another bit of propaganda from the spirit realm; Scripture tells us "there is no [governing] authority except from God, and those that exist have been instituted by God" (Romans 13:1, ESV).

It was believed that the sacred weapons forged for Ba`al by the craftsman god, Kothar-wa-Khasis (Skillful-and-Wise, or Deft-and-Clever), to defeat the sea-god Yam were kept at the temple of Hadad in Aleppo. Letters found at Mari confirm that the weapons, clubs named Yagrush (Chaser) and Aymur (Driver), were transported from Aleppo to the city of Terqa for Zimrī-Līm, the king of Mari at the time of Hammurabi (c. 1775 B.C.), and placed in the temple of Mari's chief god, Dagan.

This is fascinating on a couple levels. First, it's clear that the clubs were actual physical objects that could be brought out and displayed during ceremonies. Second, the weapons apparently had some ritual function. Whether the clubs were returned to Aleppo, we don't know, but it raises a disturbing thought: This is sheer speculation, but is it possible that these weapons, whatever they were, still exist, are in Aleppo today, and are somehow spiritually linked to the savage violence of the ongoing Syrian civil war?

Many academic papers have been written about sacred weapons of the gods and how they were used by humans who came to possess them. The clubs of Ba`al weren't the only ones. This tradition was carried over into the Christian era with only slight modifications; for example, Crusaders in the Holy Land carried what they thought was the True Cross into battle. It was lost to Salah ad-Din (Saladin) at the Battle of Hattin in 1187.

Sadly, from a historian's perspective, the temple of Hadad, which was discovered in 1929 and excavated in the late twentieth century, has been heavily damaged by the intense fighting in Aleppo that began in the summer of 2012.

Roughly eighty miles west of that city, Mount Zaphon, the home of Ba`al's palace, was known to the Greeks as Mount Kasios. As the Greek storm-god, Zeus was naturally identified with Ba`al-Hadad, and the aspect of Zeus who reigned there was accordingly known as Zeus Kasios (Jupiter Casius to the Romans). We mentioned earlier that Mount Zaphon/Kasios was the site of the epic battle between Zeus and the chaos monster, Typhon, a clear parallel with Ba`al's victory over Yam and his minion, the sea dragon Lotan.

The victory of a god over the chaos monster representing the sea, or as scholar Robert D. Miller termed it, the storm-god-slays-dragon myth, is a theme that stretches back to Sumer in the ancient Near East. The conflicts between Zeus and Typhon and Ba`al and Yam were preceded by the Hittite myth of Tarhunt and the dragon Illuyanka, the Indian myth of the god Indra's defeat of the dragon Vrtra (with a thunderbolt, naturally), and before that, the account of Marduk and Tiamat in the Babylonian creation epic, the *Enuma Elish*.

In the god lists found at Ugarit, which serve as a lexicon between Ugaritic and Akkadian, Tiamat is equated with Ba`al's nemesis, Yam. After his victory, Marduk, like Ba`al, was declared king of the gods and had a palace built in his honor.

Some scholars have observed that because no copy of the *Enuma Elish* predates the tablets containing the Baal Cycle found at Ugarit, and probably originated no more than two hundred years before the Baal Cycle, the storm-god-slays-dragon myth may well have traveled *to* Babylon from the region around Mount Zaphon and not, as is generally assumed, the other way around. And this makes a lot of sense. It's far more likely that people near the Mediterranean would envision the sea as a monstrous opponent of the gods than the inhabitants of arid central Mesopotamia.

As we mentioned earlier, the Sumerian storm-god Iškur may well have been a Semitic import. The Amorites were in contact with southern Mesopotamia from an early age. As with the storm-god himself, the account of his triumph over chaos may have traveled west to east with Amorite caravans. Of course, these tales were a PSYOP to claim the victory that Yahweh had won over Leviathan and chaos.

And there is another connection linking all these stories: We previously mentioned a letter to Zimrī-Līm, the king of Mari, confirming receipt of the weapons of Hadad at the temple of Dagan in Terqa. The king also received a message purportedly from the god himself through one of his prophets:

> Thus says Adad, I brought you back to the throne of your father, I brought you back. **The weapons with which I fought Tiamat I gave to you**. With the oil of my bitter victory I anointed you, and no one before you could stand. (*A Prophetic Letter of Adad to Zimrī-Līm* [A.1968], emphasis added)

The word translated "Tiamat," *têmtum* (a variant form of Tiamat), is a cognate of the Hebrew word *tehom*, which appears in the very second verse of the Bible:

> In the beginning, God created the heavens and the earth. The earth was without form and void, and darkness was over the face of the deep (*tehom*). And the Spirit of God was hovering over the face of the waters. (Genesis 1:1–2, ESV)

Linking the Sumerian chaos goddess, Tiamat, and "the deep" of Genesis 1:2 puts that verse in a new light. Why did the Spirit of God hover over the waters? Is it possible that Yahweh defeated a divine rebel *before* creating Adam and Eve? And having cast Tehom/Tiamat into the abyss, did His Spirit remain to guarantee the monster would stay there?

Thus, the creation of the world as recorded in Genesis is linked to the *Enuma Elish*, the Baal Cycle, and the storm-god-slays-dragon myths of ancient Anatolia and Greece, and probably the nightly contest between Set and Apep, the Indian myth referenced above, the battles between Thor and Jörmungandr, and others. Not surprisingly, scholars generally believe the biblical account was inspired by the Babylonian myth instead of the other way around—as we said, a PSYOP. Remember, the oldest written account isn't necessarily the correct one.

Why does this matter? God didn't put filler in the Bible. "All Scripture is breathed out by God and profitable for teaching, for reproof, for correction, and for training in righteousness" (2 Timothy 3:16). The victory over Leviathan was important enough that it's mentioned several times in Scripture:

> You divided the sea by your might;
> you broke the heads of the sea monsters on the waters.
> You crushed the heads of Leviathan;
> you gave him as food for the creatures of the wilderness.
> (Psalm 74:13–14, ESV)

By his power he stilled the sea;
by his understanding he shattered Rahab.
By his wind the heavens were made fair;
his hand pierced the fleeing serpent.
(Job 26:12–13, ESV)

Awake, awake, put on strength,
O arm of the LORD;
awake, as in days of old,
the generations of long ago.
Was it not you who cut Rahab in pieces,
that pierced the dragon?
Was it not you who dried up the sea,
the waters of the great deep,
who made the depths of the sea a way
for the redeemed to pass over?
(Isaiah 51:9–10, ESV)

O LORD God of hosts,
who is mighty as you are, O LORD,
with your faithfulness all around you?
You rule the raging of the sea;
when its waves rise, you still them.
You crushed Rahab like a carcass;
you scattered your enemies with your mighty arm.
The heavens are yours; the earth also is yours;
the world and all that is in it, you have founded them.
The north and the south, you have created them;
Tabor and Hermon joyously praise your name.
You have a mighty arm;
strong is your hand, high your right hand.
(Psalm 89:8–13, ESV, emphasis added)

Rahab literally means "proud one" in Hebrew. It's clear from the context of the passages above that it's linked to the sea, chaos, and evil. But unlike the lesser gods of the ancient Near East, who trembled with fear at the power and fury of the chaos sea monster, and whose champions needed the help of other gods to subdue it, Yahweh's victory was quick and unassisted.

Note also the passing references to north (*tsaphon*/Zaphon) and Hermon in Psalm 89:12 above. That's another jibe at the so-called abodes of the gods.

At the risk of repetition, a key difference between Yahweh and the small-G gods ignored by many scholars and skeptics: They are fallible, capricious, and subject to the same petty emotions that govern us mortals. None of them were credited with creating the universe and all that's in it. Yahweh is just the opposite. The Bible reveals the nature and stature of Yahweh: Creator of all things, all-powerful, all-knowing, present everywhere and every-*when*.

But at the time of the Exodus, the Bible existed only in the mind of God. The gods of Mesopotamia still ruled the nations that surrounded His chosen people. In that world, Mount Zaphon, the home of Ba`al, was so important that the Hebrew word *tsaphon* meant the compass direction north.

Now, watch this: As with Psalm 89 above, there are passages in the Old Testament that take on a startling new meaning when you substitute the proper name Zaphon for "north." For example, the famous verse that names the divine rebel Lucifer:

How art thou fallen from heaven, O Lucifer, son of the morning! how art thou cut down to the ground, which didst weaken the nations! For thou hast said in thine heart, I will ascend into heaven, I will exalt my throne above the stars of God: **I will sit also upon**

the mount of the congregation, in the sides of the north: I will ascend above the heights of the clouds; I will make myself like the Most High. (Isaiah 14:12–14, KJV, emphasis added)

Now, substitute Zaphon and watch what happens. The NIV renders verse 13 this way:

You said in your heart, "I will ascend to the heavens; I will raise my throne above the stars of God; **I will sit enthroned on the mount of assembly, on the utmost heights of Mount Zaphon.** (Isaiah 14:13, NIV, emphasis added)

The New English Translation (NET) Bible reads, "I will rule on the mountain of assembly **on the remote slopes of Zaphon.**"

That puts the prophet's words in a whole new context! Not only was Isaiah speaking to an entity, he spelled out the goal of the rebellion.

The subject of Isaiah 14 beginning at verse 12 is named in Hebrew Helel ben Shachar, Light Bringer (or Day Star) son of Dawn. (Dawn is capitalized because Šahar was the Canaanite god of the dawn, lending support to identifying the subject of the verse as a supernatural entity.) Instead of Light Bringer, Jerome rendered the Hebrew into Latin as Lucifer (*lux* + *fero*), and we English-speakers have adopted it as a personal name for Satan.

According to Isaiah, Helel's goal was nothing less than making his holy mountain, the mountain where the gods assembled, supreme in the cosmos and becoming like *elyon*—the Most High.

Which mountain? Mount Zaphon—the holy mountain of Ba`al.

———◆———

That arrogance set the tone for the next 1,500 years of Jewish and Christian history. Ba`al plagued the Israelites long before they entered the Promised Land, all the way back to the days in Egypt, through the time of Jesus. Picking up the historical narrative from Mount Sinai, Moses and his assistant Joshua led the multitudes of Israel out into the desert to begin their march to claim the Holy Land.

This led to the infamous incident where the spies, except for Joshua and Caleb, brought "a bad report of the land" to the camp and frightened the people of Israel into deciding to replace Moses and return to Egypt. Their report is the second and final place the Nephilim are mentioned in the Bible:

> "However, the people who dwell in the land are strong, and the cities are fortified and very large. And besides, **we saw the descendants of Anak there.** The Amalekites dwell in the land of the Negeb. The Hittites, the Jebusites, and the Amorites dwell in the hill country. And the Canaanites dwell by the sea, and along the Jordan."
>
> But Caleb quieted the people before Moses and said, "Let us go up at once and occupy it, for we are well able to overcome it." Then the men who had gone up with him said, "We are not able to go up against the people, for they are stronger than we are." So they brought to the people of Israel a bad report of the land that they had spied out, saying, "The land, through which we have gone to spy it out, is a land that devours its inhabitants, and **all the people that we saw in it are of great height. And there we saw the Nephilim (the sons of Anak, who come from the Nephilim), and we seemed to ourselves like grasshoppers, and so we seemed to them.**" (Numbers 13:28–33, ESV, emphasis added)

Well, you know how that turned out. As punishment, God decreed that the Israelites would wander in the desert for another forty years before entering Canaan. We've already touched on some of the supernatural encounters they had during this time, like the attacks of fiery serpents, the *nachash saraph*—as if wandering in the desert day after day wasn't punishment enough!

But there's another part of the story we never hear in church. Did you know that when Yahweh finally led Israel out from Sinai to finally begin the conquest of Canaan, He was accompanied by other gods?

God came from Teman,
and the Holy One from Mount Paran *(another name for Sinai)*.
His splendor covered the heavens,
and the earth was full of his praise. *Selah*
His brightness was like the light;
rays flashed from his hand;
and there he veiled his power.
Before him went pestilence,
and plague followed at his heels.
(Habakkuk 3:3–5, ESV, emphasis added)

You see, the words translated "pestilence" and "plague" in this passage were the names of Canaanite gods, Deber and Resheph.

Resheph was one of the more popular West Semitic gods, venerated in the city of Ebla in northern Mesopotamia as far back as the third millennium B.C. His cult extended throughout Canaan and into Egypt during the Hyksos period, where Resheph was worshipped alongside Ba`al-Set as late as the Ramesside period (thirteenth century B.C.). His followers, at least until the end of the second millennium B.C., were mostly Amorites; theophoric names (names containing the name of a deity) including Resheph as an element have been found at Mari, Terqa, and Hana on the Euphrates, as well as at Ugarit, the Canaanite coastal city-state that reached the peak of its power around the time of the Exodus.

Resheph was the lord of battle and diseases, a god with a dual nature—benevolent on one hand and dangerous on the other. He was depicted as an archer who spread war and disease with his arrows. At Ugarit, Resheph had another role—he was the gatekeeper of the netherworld. By the first millennium B.C., his cult had spread from the Levant to Cyprus and Phoenician colonies in north Africa, most prominently at Carthage.

Resheph, the Semitic god of war and plague, from an Egyptian stela dated to the 19th Dynasty (1292-1186 BC), the dynasty of Ramesses the Great.

Resheph appears in the Bible more often than you'd think. In fact, he's mentioned in the Song of Moses in Deuteronomy 32, not long after the prophet recounts how Yahweh divided the nations at Babel. Following that, Moses reminds the Israelites how, during their wanderings in the wilderness, the people forgot Yahweh and turned to worshiping the bene elohim:

> But Jeshurun grew fat, and kicked;
> you grew fat, stout, and sleek;
> then he forsook God who made him
> and scoffed at the Rock of his salvation.
> **They stirred him to jealousy with strange gods;**
> with abominations they provoked him to anger.
> **They sacrificed to demons that were no gods,**

to gods they had never known,
to new gods that had come recently,
whom your fathers had never dreaded….
"And I will heap disasters upon them;
I will spend my arrows on them;
they shall be wasted with hunger,
and devoured by **plague** [Resheph]
and **poisonous pestilence** [Qeteb, another demonic entity
known to the Canaanites];
I will send the teeth of beasts against them,
with the venom of things that crawl in the dust."
(Deuteronomy 32:15–17,23–24, ESV, emphasis added)

This is a fascinating glimpse into the supernatural realm. While they were worshipped as gods in their own right, Resheph, Deber, Qeteb, Barad (Hail), and others were still subservient to Yahweh, serving His will in dealing out punishment upon His orders.

He who dwells in the shelter of the Most High
will abide in the shadow of the Almighty.
I will say to the LORD, "My refuge and my fortress,
my God, in whom I trust."
For he will deliver you from the snare of the fowler
and from the deadly pestilence [Deber].
He will cover you with his pinions,
and under his wings you will find refuge;
his faithfulness is a shield and buckler.
You will not fear the terror of the night [Paḥad, another
demonic entity],
nor the arrow that flies by day [a reference to Resheph's
weapon, the bow],
nor the pestilence [Deber] **that stalks in darkness,**
nor the destruction [Qeteb] **that wastes at noonday.**

A thousand may fall at your side,
ten thousand at your right hand,
but it will not come near you.
You will only look with your eyes
and see the recompense of the wicked.
(Psalm 91:1–8, ESV, emphasis added)

So how does that work? Why would rebellious sons of God accompany Him and do His bidding? There are a couple of possibilities.

Remember, the Fallen lie. For all we know, Resheph and the others may have claimed to be gods with the destructive attributes of angels who remained loyal to Yahweh—a case of cosmic stolen identity. Or the prophets may have used the names of well-known gods to describe destructive aspects of God's wrath in terms people would understand— like you or me describing a storm as the work of Mother Nature. (Not that we *would*, but you get the point.)

Or the prophets simply have wanted to show their readers that Resheph and his pals were subordinate to Yahweh, inferior entities tasked with carrying out His orders whether they wanted to or not. They were not freelancing on their own authority.

That, by the way, is basically the purpose of the divine council.

As with most of the deities in the Bible, Resheph was known by other names in other cultures around the Near East. To the Babylonians, he was Nergal, a war god who became ruler of the underworld over time. Aspects of Nergal are consistent with the Hebrew understanding of Resheph's name, which is translated "flame," "lightning," "thunderbolt" (Psalm 78:48), or "burning fever" (Deuteronomy 32:24, cited above). Nergal was sometimes called *Sharappu* ("the burner"), which is related to a Hebrew word we encountered earlier—*saraph*, the root word of *seraphim*, the "burning ones."

Coincidence? Hmm.

But here's the *really* interesting bit: By the time Greece emerged as an international power, Resheph was identified with another deity who was depicted as an archer and a god of medicine and healing, but who also spread plague and disease with his arrows. And, like Resheph, this god was associated with heat and burning. One of his most important jobs was hitching the chariot team that pulled the sun across the sky every day.

Resheph was Apollo, one of the most prominent of the Olympian gods. What many Christians don't realize, because so few churches teach prophecy, is that Resheph/Apollo has a big role to play in the last days.

> And the fifth angel blew his trumpet, and I saw a star fallen from heaven to earth, and he was given the key to the shaft of the bottomless pit. He opened the shaft of the bottomless pit, and from the shaft rose smoke like the smoke of a great furnace, and the sun and the air were darkened with the smoke from the shaft. Then from the smoke came locusts on the earth, and they were given power like the power of scorpions of the earth. They were told not to harm the grass of the earth or any green plant or any tree, but only those people who do not have the seal of God on their foreheads. They were allowed to torment them for five months, but not to kill them, and their torment was like the torment of a scorpion when it stings someone. And in those days people will seek death and will not find it. They will long to die, but death will flee from them.
>
> In appearance the locusts were like horses prepared for battle: on their heads were what looked like crowns of gold; their faces were like human faces, their hair like women's hair, and their teeth like lions' teeth; they had breastplates like breastplates of iron, and the noise of their wings was like the noise of many chariots with horses rushing into battle. They have tails and stings like scorpions, and their power to hurt people for five months is in their tails. **They have as king over them the angel of the bottomless pit. His name in Hebrew is Abaddon, and in Greek he is called Apollyon.** (Revelation 9:1–11, ESV, emphasis added)

Apollyon is just another name for Apollo, and the historical evidence that Apollo is Resheph is solid. Archaeologists have found two inscriptions on the island of Cyprus that specifically identify Apollo as Resheph.[18]

The identification fits. Apollo was Resheph, god of war and pestilence, and gatekeeper of the underworld at Ugarit. He was also Nergal, the Babylonian god of war, diseases, and the underworld. The prophesied role for Apollo in the last days is completely in character with his depiction in myth and in the Bible.

Although it may be startling to see such a clear crossover between the world of Greek mythology and Scripture, it's not the only one. Spoilers— there are more coming before we're through.

And if you haven't read Tom Horn's research into the significance of the Apollonian oracle at Cumae and the Sibylline Oracles—not that they're accurate prophecies, but Christians have been deceived since the Middle Ages into believing that the Cumaean Sibyl prophesied the birth of Christ—then you should make a point of reading his book *Zenith 2016* as soon as you're done with this one. The oracle prophesied the return of "old Saturn's reign, with a new breed of men sent down from heaven." That's a reference to a return of the Golden Age of Kronos, about which we'll have more to say later.

Scholars have traced the origin of Apollo's name back through the Hittite Apaliunas, who was among the gods who guaranteed a treaty in Anatolia (Turkey) in the thirteenth century B.C.Apalunias in turn may be the older Hurrian deity Aplu, a plague god who could either bring disease or protect from it. And Aplu's name may be based on the Akkadian *Aplu Enlil* ("son of Enlil"), an epithet of the Nergal.

This is a common pattern, if you haven't noticed by now. Zeus = Ba`al = Hadad = Adad = Iškur; Venus = Aphrodite = Astarte = Ishtar = Inanna; and Apollo = Reseph = Apaliunas = Aplu = Nergal. Even though scholars tend to connect the Greek and Roman gods to an Indo-European tradition, when you pull on the threads they lead back to Sumer—and Babel.

The point here is to remind you that there's a lot more going on in the spirit realm, and more players in the game, than we usually think.

———

Let's get back to the campaign to conquer Canaan. We sidetracked and left the Israelites on the march from Mount Sinai to give you the biography of Resheph and his surprising identity as the Greek god Apollo.

Earlier, we mentioned a word that becomes extremely important as the Israelites begin to make their move on Canaan: *kherem*. It's often translated with the phrase "devoted to destruction," and it's sometimes rendered "under the ban." It basically means something that is taboo—as in "touch this and you die."

As it happened, many of the people living in the land that Yahweh chose for His people were *kherem*. But not all of them, and He was very specific about who was who. God used several different verbs to describe the conquest of Canaan: *Garash, yarash*, and the aforementioned *kherem*.

Garash means "to drive out." For example:

And I will send hornets before you, which shall **drive out** the Hivites, the Canaanites, and the Hittites from before you. I will not **drive them out** from before you in one year, lest the land become desolate and the wild beasts multiply against you. Little by little I will **drive them out** from before you, until you have increased and possess the land. And I will set your border from the Red Sea to the Sea of the Philistines, and from the wilderness to the Euphrates, for I will give the inhabitants of the land into your hand, and you shall **drive them out** before you. (Exodus 23:28–31, ESV, emphasis added)

Similarly, *yarash* means "to dispossess, drive out":

For I will **cast out** nations before you and enlarge your borders; no one shall covet your land, when you go up to appear before the LORD your God three times in the year. (Exodus 34:24, ESV, emphasis added)

And Moses sent to spy out Jazer, and they captured its villages and **dispossessed** the Amorites who were there. (Numbers 21:32, ESV, emphasis added)

Now these are the kings of the land whom the people of Israel defeated and **took possession** of their land beyond the Jordan toward the sunrise. (Joshua 12:1, ESV, emphasis added)

Kherem, though, was a completely different ball of wax. It meant total extermination. Take no spoil and no prisoners. Offer no mercy and no deals.

When the LORD your God brings you into the land that you are entering to take possession of it, and clears away many nations before you, the Hittites, the Girgashites, the Amorites, the Canaanites, the Perizzites, the Hivites, and the Jebusites, seven nations more numerous and mightier than yourselves, and when the LORD your God gives them over to you, and you defeat them, then you must **devote them to complete destruction**. You shall make no covenant with them and show no mercy to them. You shall not intermarry with them, giving your daughters to their sons or taking their daughters for your sons, for they would turn away your sons from following me, to serve other gods. Then the anger of the LORD would be kindled against you, and he would destroy you quickly. But thus shall you deal with them: you shall break down their altars and dash in pieces their pillars and chop down their Asherim and burn their carved images with fire. (Deuteronomy 7:1–5, ESV, emphasis added)

That sounds harsh. It is. Remember, though, these people were sold out completely to the rebellious bene elohim who were themselves under a death sentence pronounced by Yahweh in the divine council: "You are gods, sons of the Most High, all of you; nevertheless, like men you shall die, and fall like any prince" (Psalm 82:6–7, ESV).

The gods were desperate, and they still are. They're doomed and they know it. For thousands of years, their goal has been to overthrow the Most High, but if they can't manage that they'll take as many of us with them as they can into the lake of fire.

So who was targeted for the ban? Well, not Israel's cousins. Edom, Moab, and Ammon, nations on the east side of the Jordan River, were off limits. Edom was founded by Jacob's brother Esau, while Moab and Ammon were the descendants of Abraham's nephew Lot. God told Moses in Deuteronomy 2 that He had given those lands to the Edomites, Moabites, and Ammonites as a possession.

Interestingly, it appears that God used those nations to clear the way for Israel by eliminating tribes of giants who had formerly occupied those lands:

> Then the LORD said to me, "You have been traveling around this mountain country long enough. Turn northward and command the people, 'You are about to pass through the territory of your brothers, the people of Esau, who live in Seir; and they will be afraid of you. So be very careful. Do not contend with them, for I will not give you any of their land, no, not so much as for the sole of the foot to tread on, because I have given Mount Seir to Esau as a possession.'"…
>
> And we turned and went in the direction of the wilderness of Moab. And the LORD said to me, "Do not harass Moab or contend with them in battle, for I will not give you any of their land for a possession, because I have given Ar to the people of Lot for a possession." (**The Emim formerly lived there, a people great and many, and tall as the Anakim. Like the Anakim they are also counted as Rephaim, but the Moabites call them Emim. The Horites also lived in Seir formerly, but the people of Esau dispossessed them and destroyed them from before them and**

settled in their place, as Israel did to the land of their possession, which the LORD gave to them.)…

"Today you are to cross the border of Moab at Ar. And when you approach the territory of the people of Ammon, do not harass them or contend with them, for I will not give you any of the land of the people of Ammon as a possession, because I have given it to the sons of Lot for a possession." (**It is also counted as a land of Rephaim. Rephaim formerly lived there—but the Ammonites call them Zamzummim—a people great and many, and tall as the Anakim; but the LORD destroyed them before the Ammonites, and they dispossessed them and settled in their place, as he did for the people of Esau, who live in Seir, when he destroyed the Horites before them and they dispossessed them and settled in their place even to this day. As for the Avvim, who lived in villages as far as Gaza, the Caphtorim, who came from Caphtor, destroyed them and settled in their place.**) (Deuteronomy 2:2–5, 8b–12, 18–23, ESV, emphasis added)

There isn't much we know about the Horites and the Avvim. They aren't well attested outside the Bible, although a people mentioned in Egyptian texts who lived in the southern part of Canaan called the Khar may be the biblical Horites. Certainly, Mount Hor in Edom was named for the Horites, or vice versa, and the Edomites themselves were descendants of the Horites since Esau took as a wife the daughter of a Horite chief, Zibeon.

The Rephaim, however, are another story. In fact, it's a story that will take some time to unpack, so we're going to hold it back for a bit to develop some background first. You've guessed from the context that they were tall, but there's a lot more to it than that.

Now, even though the cousins of the Israelites had done some of the heavy lifting east of the Jordan, there were still two kingdoms on Yahweh's hit list—Heshbon, led by the Amorite king Sihon, and Bashan, which was ruled by Og, who was called the last of the Rephaim.

This is significant. The land Yahweh had reserved for the Israelites was Canaan, usually understood to mean the territory west of the Jordan. So why would God give Moses the green light to devote to destruction these two kingdoms bordering on Moab and Ammon?

Good question. We'll answer that one in the next section, too. The point is that both Sihon of Heshbon and Og of Bashan were targeted and their small kingdoms utterly destroyed—*kherem*.

But Sihon the king of Heshbon would not let us pass by him, for the LORD your God hardened his spirit and made his heart obstinate, that he might give him into your hand, as he is this day. And the LORD said to me, "Behold, I have begun to give Sihon and his land over to you. Begin to take possession, that you may occupy his land." Then Sihon came out against us, he and all his people, to battle at Jahaz. And the LORD our God gave him over to us, and we defeated him and his sons and all his people. And we captured all his cities at that time and **devoted to destruction** every city, men, women, and children. We left no survivors.... Then we turned and went up the way to Bashan. And Og the king of Bashan came out against us, he and all his people, to battle at Edrei. But the LORD said to me, "Do not fear him, for I have given him and all his people and his land into your hand. And you shall do to him as you did to Sihon the king of the Amorites, who lived at Heshbon." So the LORD our God gave into our hand Og also, the king of Bashan, and all his people, and we struck him down until he had no survivor left. And we took all his cities at that time—there was not a city that we did not take from them— sixty cities, the whole region of Argob, the kingdom of Og in Bashan. All these were cities fortified with high walls, gates, and bars, besides very many unwalled villages. And we **devoted them to destruction**, as we did to Sihon the king of Heshbon, devoting to destruction every city, men, women, and children. But all the livestock and the spoil of the cities we took as our plunder.

So we took the land at that time out of the hand of the two kings of the Amorites who were beyond the Jordan, from the Valley of the Arnon to Mount Hermon (the Sidonians call Hermon Sirion, while the Amorites call it Senir), all the cities of the tableland and all Gilead and all Bashan, as far as Salecah and Edrei, cities of the kingdom of Og in Bashan. (Deuteronomy 2:30–34, 3:1–10, ESV, emphasis added)

Having struck the first blows in their war of conquest, the Israelites turned their eyes across the Jordan. Moses wasn't allowed to go as punishment for his disobedience at the rock at Meribah (see Numbers 20). The task of leading the conquest was put in the hands of Joshua, who had a plan: Divide and conquer.

A careful analysis of the book of Joshua reveals that his strategy—rather, God's strategy—was aimed at the Anakim the spies reported seeing back in Numbers 13. The Anakim, descendants of the Nephilim, were concentrated in the hill country. The invasion of the Israelites was divided into two phases, first a southern campaign followed by a campaign in the north, beginning at the Amorite stronghold on the Jordan River, Jericho.

The battle against Og at Edrei was in the northeast of Canaan, due east of the Sea of Galilee. Og's other chief city, Ashtaroth, was near the Golan Heights. To open the invasion of Canaan at Jericho, Joshua first had to lead Israel south, back the way it had come. From a logistical standpoint, they could have simply begun their invasion in the north against the powerful city-state of Hazor. Instead, they backtracked to begin the conquest by taking on the oldest human settlement in the Levant, Jericho.

Why Jericho? Our friend Dr. Michael Lake put forward an interesting theory in his recent book, *The Sheeriyth Imperative*, linking Jericho to the Nephilim and demonic spiritual practices. Dr. Lake suggests that the ancient city was a power center for the Amorites and Canaanites of the Holy Land.

Actually, we agree. There's no question that Yahweh had a reason for leading Joshua to begin the conquest there. He never does anything without a reason.

Now, this next bit of information may be coincidental, but you know we don't go much for coincidence theories around here. Jericho is one of the sites in the Near East where cranial deformation was practiced in pre-Flood days—in fact, the evidence suggests it was one of the first places it was practiced. After the Flood, the people at Jericho practiced a weird veneration of human skulls. Remains of the dead were defleshed and buried under the floors of their homes—except for the skulls. The skulls were filled with plaster and decorated to look *sort of* life-like and then apparently kept on display in the home.

Why? Nobody is really sure, but we'll speculate on that before the end of this book.

Before the attack on Jericho, which is the part of the story everybody knows, there was another spiritual poke in the eye of the rebellious bene elohim.

> The LORD said to Joshua, "Today I will begin to exalt you in the sight of all Israel, that they may know that, as I was with Moses, so I will be with you. And as for you, command the priests who bear the ark of the covenant, 'When you come to the brink of the waters of the Jordan, you shall stand still in the Jordan.'"
>
> And Joshua said to the people of Israel, "Come here and listen to the words of the LORD your God." And Joshua said, "Here is how you shall know that the living God is among you and that he will without fail drive out from before you the Canaanites, the Hittites, the Hivites, the Perizzites, the Girgashites, the Amorites, and the Jebusites. Behold, the ark of the covenant of the Lord of all the earth is passing over before you into the Jordan. Now therefore take twelve men from the tribes of Israel, from each tribe a man. And when the soles of the feet of the priests bearing the ark of the LORD, the Lord of all the earth, shall rest in the waters of

the Jordan, the waters of the Jordan shall be cut off from flowing, and the waters coming down from above shall stand in one heap."

So when the people set out from their tents to pass over the Jordan with the priests bearing the ark of the covenant before the people, and as soon as those bearing the ark had come as far as the Jordan, and the feet of the priests bearing the ark were dipped in the brink of the water (now the Jordan overflows all its banks throughout the time of harvest), the waters coming down from above stood and rose up in a heap very far away, at Adam, the city that is beside Zarethan, and those flowing down toward the Sea of the Arabah, the Salt Sea, were completely cut off. And the people passed over opposite Jericho. Now the priests bearing the ark of the covenant of the LORD stood firmly on dry ground in the midst of the Jordan, and all Israel was passing over on dry ground until all the nation finished passing over the Jordan. (Joshua 3:7–17, ESV)

Another parting of the waters. Cool, right? But just like the crossing of the Red Sea, it was more than just a convenient route for Israel to get into Canaan.

In the Baal Cycle, the storm-god uses his magic clubs Chaser and Driver to defeat the chaos god of the sea, Yam. But Yam is referred to in the tale as Prince Yam (Prince Sea) and Judge Nahar—Judge *River*.

Once again, Yahweh sent a clear signal to all parties concerned, human and otherwise, that He was sovereign, even over territory that was supposed to be the domain of Ba`al-Hadad, the chief god of the people who lived in the land that would become Israel. To put an exclamation point on the message, the Israelites were commanded to build a monument on the west side of the river from stones taken "out of the midst of the Jordan, from the very place where the priests' feet stood firmly" (Joshua 4:3). And then, before the water began to flow again, Joshua set up twelve stones on the bed of the river where the priests had stood. Call that an extra dash of salt in the spiritual wound.

Well, the message was received loud and clear.

As soon as all the kings of the Amorites who were beyond the Jordan to the west, and all the kings of the Canaanites who were by the sea, heard that the LORD had dried up the waters of the Jordan for the people of Israel until they had crossed over, their hearts melted and there was no longer any spirit in them because of the people of Israel. (Joshua 5:1, ESV)

The evidence of history supports the theory that the Amorite connection to Jericho—remember the iniquity of the Amorites?—is what led Yahweh to start the invasion there.

As we mentioned, Jericho was probably the oldest settlement in the Levant. Amorites moved into the area around 2300 B.C. and apparently renamed the old city for one of their tribes, the Yarihû. The Yarihû were part of a bigger confederation within the Amorite people, the Binū-Yamina, or Bene Yamina—"sons of the right hand" (southerners, since south is on the right as one faced the rising sun). The name of the Yarihû tribe appears to have come from the name of the moon-god, Yarikh, from the Semitic word for moon, *yareah*.

Jericho was a well-known center of the moon-god's cult. But it was small potatoes compared to the two major cities devoted to the moon-god in the ancient Near East.

The lunar deity was the patron of the Sumerian city of Ur, where he was called Nanna (Sumerian) or Sîn (Akkadian). The great ziggurat of Ur was dedicated to Nanna by Ur-Nammu, founder of the Third Dynasty of Ur.

Get this: The Amorites tribes of the Binū Yamina confederation would meet at the *other* major center of the moon-god cult to agree on treaties—the temple of Sîn in Harran.

And because you've been paying attention, you remember that it was Harran that Abraham began the journey that led to the Israelites marching around the walls of Jericho 470 years later. (And it was

Sumerian Ur that Christians have incorrectly identified as Abram's original family home.)

Coincidence?

So the first thrust of the spear into Canaan was directed at a city that was a base for a tribe of Amorites who venerated the moon-god. But as with the crossings of the Red Sea and the Jordan River, the attack on Jericho was another example of Yahweh demonstrating His authority over the rebellious gods of the nations.

Dr. Chuck Missler, as always, offered intriguing insights into the attack. He pointed out that the Israelites began the conquest of Canaan by violating the Torah.[19] First, the Levites, who weren't ever supposed to go to war, led the Israelite soldiers in their daily procession around Jericho. Then instead of resting on the Sabbath, the Israelites did seven times as much work!

The late Dr. David Livingston, who directed Associates for Biblical Research for twenty-five years, was thinking along the same lines we're investigating in this book. (Although, to be clear, he interpreted the bene elohim of Genesis 6 not as rebellious angels, but as despotic kings who claimed divinity.) Dr. Livingston draws a connection, noticed by secular scholars, between a Canaanite myth called *The Legend of Keret* and the tactics ordered by Yahweh for the attack on Jericho.[20]

In the legend, King Keret, described as a son of the great god El, suffers a series of misfortunes that leaves him without a wife or heir. Hearing his tearful prayers, Keret is visited in a dream by El. El instructs Keret to make war on the kingdom of Udum, where he will find a wife to give him a son. He's to hold a feast for his people and then set out for Udum—men of war first, followed by the people of Keret's kingdom, and then the trumpeters last. All the people are warned to stay quiet until the last day. Two six-day periods are recorded with a noisy climax on the seventh days. In a nutshell, Keret gets his wife, takes her home and has a son.

The parallels to the assault on Jericho are obvious. The Israelites began their assault on Jericho by celebrating the Passover. And then Joshua received orders from Yahweh-but-not-Yahweh—another Christophany.

When Joshua was by Jericho, he lifted up his eyes and looked, and behold, a man was standing before him with his drawn sword in his hand. And Joshua went to him and said to him, "Are you for us, or for our adversaries?" And he said, "No; but I am the commander of the army of the LORD. Now I have come." And Joshua fell on his face to the earth and worshiped and said to him, "What does my lord say to his servant?"

And the commander of the LORD's army said to Joshua, "Take off your sandals from your feet, for the place where you are standing is holy." And Joshua did so.

Now Jericho was shut up inside and outside because of the people of Israel. None went out, and none came in. And the LORD said to Joshua, "See, I have given Jericho into your hand, with its king and mighty men of valor. You shall march around the city, all the men of war going around the city once. Thus shall you do for six days. Seven priests shall bear seven trumpets of rams' horns before the ark. On the seventh day you shall march around the city seven times, and the priests shall blow the trumpets. And when they make a long blast with the ram's horn, when you hear the sound of the trumpet, then all the people shall shout with a great shout, and the wall of the city will fall down flat, and the people shall go up, everyone straight before him." (Joshua 5:13–6:5, ESV)

Note that Joshua was speaking to the commander of the Lord's army in chapter 5, and suddenly, in Joshua 6:2, the visitor is identified as Yahweh. As with the burning bush passage from Exodus, this is not a mistake. Just as Jesus is Yahweh, and the angel of Yahweh is Yahweh, the commander of the Lord's army was likewise the second member of the Trinity—as if we needed a reminder that the coming battle was going to be miraculous.

It's worth noting that Joshua was visited by another Joshua—Yeshua, Jesus.

Predictably, secular scholars, treating the account in Joshua as literature instead of history, generally believe that if there is any link between *The*

Legend of Keret and the book of Joshua, it's because the author of the biblical account copied the Ugaritic tale, which itself may have been adapted from an older Mesopotamian myth. The differences, of course, are the key: The priests of Yahweh led the procession of Israelites around Jericho while King Keret's warriors led the march on Udum. The victory was due to the Lord, not through the force of arms.

And just as Yahweh showed His mastery over Ba`al on the storm-god's home turf at the Red Sea and the Jordan River, the worshipers of the moon-god Yarikh learned the hard way that there is no home-field advantage when you're up against Yahweh.

This battle is filled with signs that Yahweh was working supernaturally for the Israelites. To be sure that His people were spiritually prepared, and to assure that they would trust in Him and not their own strength, Yahweh directed that the men be circumcised before the attack on Jericho. Now, the Bible does tell us Israel waited until the soldiers had healed to attack the city, but you can just imagine the thoughts in the minds of the people: "What if the Amorites attack while our fighting men are still, uh, sore?" (Come on—*flint knives!*)

This emphasizes that something was at work in the spirit realm. Knowing that an army was outside the walls, don't you think the king of Jericho had at least a couple pair of eyes on the Israelite camp? And wouldn't he have put out a call for backup to the other Amorite city-states in Canaan? After all, a mass circumcision would have been hard to miss. There were forty thousand Israelite men of military age, and none of them had been circumcised during the forty years in the desert. No wonder they named the place Gibeath-haaraloth—the Hill of Foreskins!

But even though the Israelite army was incapacitated, the Amorites didn't attack. The people of the land were literally paralyzed with fear. That's spiritual.

And the battle itself was another supernatural jab by Yahweh at the Fallen. Remember, the Israelites celebrated the Passover before going to war. Joshua 5:10 tells us that it was kept "on the fourteenth day of the month in the evening on the plains of Jericho." The next day began, as it does to this day in the Hebrew calendar, at sundown. And to this day, Passover is kept on the fifteenth of Nisan, a date that floats through the secular calendar because the feasts are based on a lunar calendar. Because the date of the attack is linked to the Passover, it would have been in the spring, probably early May in 1406 B.C.

So putting these clues together, we find something that *couldn't* have been a coincidence: Yahweh ordered Israel to attack a people and city devoted to the moon-god *on the first full moon of the new year.*[21]

Coincidence?

There is probably a lot more to it than that, but a thorough study of the Babylonian *akitu* festival, a new year ritual performed every spring that included the "sacred marriage" of the god Marduk and his consort, will have to wait for another book.

———

After Jericho and a misstep at Ai, the only battle the Israelites lost during the Conquest, because an Israelite named Achan took "devoted things" (*kherem*) from Jericho, the Israelites swept through the southern hill country of Judah and the Negev. It began when Adoni-zedek, the Amorite king of Jerusalem, called on fellow kings in Lachish, Hebron, Jarmuth, and Eglon to join him in an attack on Gibeon. The Gibeonites had just tricked Joshua into an alliance and the Amorites were panicked: Gibeon was bigger than Ai, which the Israelites had just leveled, and their fighting men had a good reputation in Canaan. Obviously, the leaders of Gibeon knew what had happened at Jericho and Ai and they didn't want to be next on the "devoted to destruction" list.

The Amorite kings were also shaken because Gibeon was a strategic prize. It commanded the central highlands of Judah. Losing it to Israel cut off any help from the north.

What Adoni-zedek and his allies didn't realize was that they were playing right into Yahweh's hands. Instead of a war of attrition, besieging one fortified city at a time, Israel could take on all the military forces of southern Canaan in one battle.

And that's exactly what happened: Not only was it a smashing victory, the Lord again demonstrated His authority over all Creation *and* the gods who'd enthralled the Amorites by—get this—stopping the rotation of the earth.

> At that time Joshua spoke to the LORD in the day when the LORD gave the Amorites over to the sons of Israel, and he said in the sight of Israel,
>   "Sun, stand still at Gibeon,
>   and moon, in the Valley of Aijalon."
> And the sun stood still, and the moon stopped, until the nation took vengeance on their enemies.
>   Is this not written in the Book of Jashar? The sun stopped in the midst of heaven and did not hurry to set for about a whole day. (Joshua 10:12–13, ESV)

We may be stretching a point, but it's known that the Semitic sun-god Šamaš, the moon-god Sîn, and Ba`al-Hadad formed a sort of divine triad in some parts of Mesopotamia, representing the three things essential to life—sun, moon, and life-bringing rain. (In other settings, Ishtar/Inanna, represented by the planet Venus, joined Šamaš and Sîn in the triad.) It's possible that Yahweh went for the trifecta in the Valley of Aijalon, showing His power over three Amorite gods at once by literally stopping them in their tracks.

Chuck Missler and others have speculated that Joshua's Long Day might have been caused by a near pass of the earth by the planet Mars. Whatever the science behind it, there's no question the hand of God was in it. Thus, the cities of Makkedah, Libnah, Lachish, Gezer, Eglon, Hebron, and Debir were put to the sword, and all of them were *kherem*.

So Joshua struck the whole land, the hill country and the Negeb and the lowland and the slopes, and all their kings. He left none remaining, but **devoted to destruction** all that breathed, just as the LORD God of Israel commanded. (Joshua 10:40, ESV, emphasis added)

While it's fascinating to think about another planet buzzing the earth and the other implications it might have for world history, not to mention what it might mean for the future (we're looking at *you*, Planet X), it overlooks a simple explanation: Joshua's Long Day capped Yahweh's victory over the moon-god Yarikh/Sîn by keeping the sun in the sky, and the moon out of it, for an entire day.

And then the northern campaign began. Hazor, the most powerful city-state in Canaan, pulled together a coalition from the smaller cities of the northern hill country. Despite the coalition's advantage in numbers, and the tactical superiority of bringing chariots to the battle, Joshua and the Israelites dealt them a crushing defeat. The result? Pretty much what you expect.

Joshua turned back at that time and captured Hazor and struck its king with the sword, for Hazor formerly was the head of all those kingdoms. And they struck with the sword all who were in it, **devoting them to destruction**; there was none left that breathed. And he burned Hazor with fire. And all the cities of those kings, and all their kings, Joshua captured, and struck them with the edge of the sword, **devoting them to destruction**, just as Moses the servant of the LORD had commanded. (Joshua 11:10–12, ESV, emphasis added)

At this point, you might be asking yourself, "Was all this devoting to destruction really necessary?" That's understandable, but don't go too far down that road. Yahweh is not a genocidal sociopath. There were reasons for ordering the Israelites to do what they did.

First of all, the war appears to have targeted the strongholds of the Anakim, the descendants of the giants—the Nephilim. Hebron, for example, was on the list of cities targeted for *kherem*. Apparently, it was a stronghold of the giants. Remember back to the report of the spies, forty years before the conquest:

> So they went up and spied out the land from the wilderness of Zin to Rehob, near Lebo-hamath. They went up into the Negeb and came to Hebron. Ahiman, Sheshai, and Talmai, **the descendants of Anak, were there.** (Hebron was built seven years before Zoan in Egypt.)…
>
> However, the people who dwell in the land are strong, and the cities are fortified and very large. **And besides, we saw the descendants of Anak there.** The Amalekites dwell in the land of the Negeb. The Hittites, the Jebusites, and **the Amorites dwell in the hill country.** And the Canaanites dwell by the sea, and along the Jordan.…
>
> **And there we saw the Nephilim (the sons of Anak, who come from the Nephilim), and we seemed to ourselves like grasshoppers, and so we seemed to them.** (Numbers 13:21–22, 28–29, 33, ESV, emphasis added)

We learn in Joshua 11:21 that the Anakim and the Amorites were concentrated in the hill country of Judah and Israel, which makes up most of Israel between the coastal plain and the Jordan River valley. So what do we see in the book of Joshua? A military campaign targeting cities in the hill country. It's no surprise, then, that the king of Hazor, in the hill country of Galilee, in the north end of the Holy Land, called on other cities in the highlands for soldiers to fight the Israelite menace.

The Anakim weren't a specific nation in the land of Canaan. The biblical account portrays them as a bloodline considered an abomination by Yahweh, the descendants of the Nephilim. And their location alongside the Amorites in the hill country is significant. The Anakim were, at the very least, allies of the Amorites if not a subgroup.

Jewish literature from the Second Temple period makes it clear that the sin of the Watchers was more than just lusting after human women and foisting the Nephilim giants on humanity. The technology and spiritual practices the Watchers passed along—divination, sorcery, necromancy, warcraft, etc.—were largely responsible for their incarceration in Tartarus until the judgment.

The Bible also reports that some of the Nephilim were still around after the Flood. How, if all flesh on land was killed? We don't know for sure, but multiple incursions by the bene elohim is hinted at in Genesis 6:4 ("when" could also be translated "whenever").

The Anakim were their descendants. And you know what? We have proof from outside the Bible that the Anakim really existed.

During the second millennium B.C., the Egyptians had a practice of inscribing curses against enemies on clay figurines, bowls, or blocks of clay or stone, which were then smashed. It was their version of sticking pins into voodoo dolls. Several clay shards from these Execration Texts dated to the twelfth or thirteenth dynasties, the nineteenth or eighteenth century B.C. (just before Jacob and his family emigrated to Egypt), mention rulers of *Iy-anaq*—the Anakim.

The Ruler of Iy'anaq, Erum, and all the retainers who are with him; the Ruler of Iy'anaq, Abi-yamimu and all the retainers who are with him; the Ruler of Iy'anaq 'Akirum and the retainers who are with him.[22]

Not only does this confirm that Anakim were in the land at the time of Abraham and Isaac (and therefore it's not a stretch to believe they were still around at the time of the Exodus), but they were enough of a nuisance for the Egyptians to sic their gods on them!

Further, a letter from an Egyptian scribe named Hori to another named Amenemope that's been preserved describes a mountain pass near Megiddo as "infested with Shasu," people apparently known for their size. The Shasu were cattle nomads who lived throughout the Levant.

According to Hori the scribe, "some of them are of four cubits or of five cubits, from head to foot." Since the Egyptian royal cubit was a little over twenty and one-half inches, that would make those Shasu between 6'8" and 8'6".[23]

An Egyptian inscription commemorating the Battle of Kadesh, where the forces of Ramesses the Great went toe to toe with those of the Hittite king Muwatalli II in 1274 B.C. (about 125 years after the Conquest), shows Egyptian soldiers using enhanced interrogation techniques on a pair of Shasu spies. (In plain English, the Egyptians were beating them up.) Get this: The Shasu prisoners are as tall *on their knees* as the Egyptian soldiers.

Shasu spies being beaten by Egyptians (detail from the Battle of Kadesh wall carving).

Now, it's one thing for the Egyptians to show their kings as superhumanly big. They did that all the time. So when the Egyptians did it for enemies, it wasn't because they were trying to make them look good. The Shasu were just exceptionally big. Again, this supports the biblical account of giants in the land.

So, given that the Anakim were descendants of the rebellious Watchers, it's no surprise that Joshua made ridding the land of them the goal of the war:

And Joshua came at that time and cut off the Anakim from the hill country, from Hebron, from Debir, from Anab, and from all the hill country of Judah, and from all the hill country of Israel. Joshua devoted them to destruction with their cities. **There was none of the Anakim left in the land of the people of Israel. Only in Gaza, in Gath, and in Ashdod did some remain.** So Joshua took the whole land, according to all that the LORD had spoken to Moses. And Joshua gave it for an inheritance to Israel according to their tribal allotments. And the land had rest from war. (Joshua 11:21–23, ESV, emphasis added)

Only in the future cities of the Philistines did any of the Anakim remain. Goliath, the giant who stared down the entire Israelite army under King Saul, was from Gath, as were the other four giants killed by David and his mighty men (2 Samuel 21:15–22).

Joshua's summary account of the war focused on the Anakim, not the Amorites, Canaanites, or other -ites. As with everything in the Bible, that's not by accident. The Anakim, descendants of the Nephilim, were sold out to the Fallen in more ways than one, and their presence was an existential threat to Yahweh's chosen people and, most importantly, the future Messiah.

It was not an accident, either, that the Anakim were in the land waiting for the Israelites. Remember, they'd known for more than four hundred years that the Israelites were coming, thanks to the small-G gods they served. They just hadn't planned on Yahweh's active participation in the fight.

But there is still a question that we need to address about the conquest, and it has to do with two of the people groups Yahweh pushed out of the path of the Israelites, the Canaanites and the Amorites.

The timing of the Exodus has always been a puzzle. Like everything God does, it happened at exactly the time He chose and for a specific reason. Remember back to the covenant Yahweh made with Abraham in Genesis 15: The patriarch was told his descendants would return from

Egypt "in the fourth generation, for the iniquity of the Amorites is not yet complete."

Yes, we're finally going to get to that. Thank you for your patience.

———

The obvious questions that come to mind: Who were the Amorites, and why was the timing of the Exodus linked to their iniquity? What *was* their iniquity? What could they have done that was so bad that God made it a signpost on the road to Revelation? Whatever it was, the evil of the Amorites was legendary among the Jews:

> And the LORD said by his servants the prophets,
> "Because Manasseh king of Judah has committed these abominations and has done things **more evil than all that the Amorites did**, who were before him, and has made Judah also to sin with his idols, therefore thus says the LORD, the God of Israel: Behold, I am bringing upon Jerusalem and Judah such disaster that the ears of everyone who hears of it will tingle." (2 Kings 21:10–12, ESV, emphasis added)

Manasseh was king of Judah about seven hundred years after the Exodus, nearly 1,200 years after Abraham was first called from Ura, near Harran. Whatever the Amorites did, it was *bad*.

The Amorites were incredibly resilient. They hung around, and dominated for a long while, a part of the world where people have been warring since the time of Nimrod. They were a Semitic speaking people who occupied nearly the entire Near East during the first half of the second millennium B.C. According to the Bible, the Amorites descend from Noah's son, Ham, by way of Canaan. However, even though Ham is considered the progenitor of various African races, Egyptian artists usually represented Amorites with fair skin, light hair, and blue eyes.

The Amorites were first mentioned in Mesopotamian records around

2400 B.C., or just before Sargon the Great turned Akkad from a city-state into an empire. They were known to the city-dwelling Sumerians as the MAR.TU, who considered them savage, uncouth, and generally unpleasant.

> The MAR.TU who know no grain.... The MAR.TU who know no house nor town, the boors of the mountains.... The MAR. TU who digs up truffles.... who does not bend his knees (to cultivate the land), who eats raw meat, who has no house during his lifetime, who is not buried after death.[24]

Scholars haven't reached a consensus over just where the Amorites came from. The Ebla texts refer to an Amorite LU.GAL, or king, named Amuti, in the 2300s B.C. The Amorite kingdom, MAR.TU[ki] in Sumerian, seems to have been centered in Syria around Jebel Bishri, a mountain on the west bank of the Euphrates about thirty miles west of Deir ez-Zor. The mountain of the Amorites, called Bašar back in the day, was the site of a military victory led by the Akkadian king Narām-Sîn, grandson of Sargon the Great, over a coalition of Amorites led by Rish-Adad, the lord of a small city called Apishal.

Evidence uncovered by a team of Finnish researchers who began work at Jebel Bishri in 2000 indicates that the main urban center of MAR. TU[ki] in the third millennium B.C. was Tuttul, a city on the Euphrates near the modern city of Raqqa. Tuttul was settled by the twenty-sixth century B.C. and was sacred to Dagan, chief god of the local pantheon. A second temple of Dagan was built later at Terqa, south of Tuttul on the Euphrates about halfway to Mari. That's the temple we mentioned earlier that received the sacred clubs of Ba`al-Hadad. Dagan was also the god of the L m dynasty that ruled the kingdom of Mari, farther downriver near where the Euphrates crosses from Syria into Iraq.

This leads to a bigger question: What exactly was the religion of the Amorites? Might that shed some light on why Yahweh called them out when He made the covenant with Abraham?

Scholars for many years thought the Amorites worshiped a national god named Amurru, which was the Akkadian name by which the Amorites themselves were called. That opinion has changed as scholars realized there are absolutely no Amorite names that include Amurru as a theophoric element. In other words, if they worshiped Amurru, at least *one* of them would have a name that means "Amurru helps." "Amurru listens." "Amurru is my strength," or something like that. But that's not the case.

Judging by personal names in the earliest records, it looks as though there were originally only two main gods of the Amorites—the moon-god, Ereah (Yarikh, hence the name of the tribe and the moon-god's cult center, Jericho), and "the" god, El.

That should interest anyone who's read any of the Old Testament. Besides being the name of the chief god of the Canaanite pantheon, El was one of the names the prophets applied to Yahweh. Remember that El was the *only* name by which Yahweh was known to Abraham, Isaac, and Jacob. That's significant because they lived during the time the Amorites controlled the Near East.

It would appear—and this is just speculation by your author—that the Fallen took a binary view of the cosmos, good (El/Yahweh) vs. evil (the moon-god?), and twisted it—a classic PSYOP. The best lies are grown from a kernel of truth. By the time Joshua led the Israelites into Canaan, El had been transformed into an almost buffoonish character. In one myth, the high god of the Amorites got so drunk at a feast that he wallowed in his own filth and urine.

That was *not* Yahweh.

Meanwhile, the moon-god, while not as well-known as Satan, Ba`al, or Marduk, enjoyed a long run as one of the most influential gods of the Near East. After the defeat at Jericho, Yarikh (Sîn to the Akkadians, Nanna to the Sumerians) eventually reemerged as the chief god of Babylon nearly a thousand years later under king Nabonidus (reigned 556–539 B.C.). And one can make a strong case that the moon-god today leads what may become the largest religion on the planet within the next half-century. More about that later.

However, as they gained political control over regions previously governed by the Akkadians and Sumerians after 2000 B.C., the Amorites began adopting the gods of their subjects. A classic example is the Amorite chief, Šamši-Adad I. His father, Ila-kabkabu (possibly "El is my star"), had been the king of Terqa. After being forced to flee to Babylon during the reign of Narām-Sîn of Akkad, Šamši-Adad eventually returned home, overthrew Narām-Sîn's successor, and became the first Amorite king of Assyria.

Šamši-Adad, whose name includes that of the storm-god of Aleppo, named the son he placed on the throne of Mari, whose territory bordered on Yamḥad (Aleppo), Yasmah-Adad. The son who governed the Akkadian part of his realm was named Išme-Dagan. Both names mean "(deity) hears," but presumably Adad and Dagan were more acceptable to the subjects of their respective parts of the realm than Amorite gods like Ilu (El) or Ereah (Yarikh).

We get a hint of conflict, either between human kings or the gods they served, in the transfer of the sacred weapons of Hadad to the temple of Dagan when Zimrī-Līm recovered his throne (c. 1775 B.C.). The event is interpreted by some scholars as a political move by the king of Yamḥad to assert authority over Mari. There may be some truth to that. The subtext of the message from the storm-god to the king of Mari was that since Hadad restored the fortunes of the Līm family, Hadad could tear it down again—presumably over the wishes of Dagan, who obviously hadn't been powerful enough to keep Zimrī-Līm's father on his throne in the first place.

On the other hand, viewing the message through a supernatural lens, it might have been a message from Ba`al-Hadad to Dagan—a reminder to the old god that there was a new marshal in town.

In case you're wondering, the theophoric element "Līm" used by several rulers in northwest Mesopotamia seems to be based on the Akkadian *līm ilāni*, the "thousand gods." The *līm ilāni* were invoked as witnesses to treaties, and scholars think the term probably means the entire pantheon of gods—in other words, the divine assembly.

One wonders whether there is any relationship between the *līm ilāni* and the spirits inside the Gadarene demoniac in Mark 5:9 and Luke 8:30: "My name is Legion, for we are many." But I digress.

Since various factions within the broad scope of Amorite ethnicity followed different gods, the bene elohim may have been jockeying for power between themselves. That's sheer speculation, but it seems consistent with the evidence of history. On the other hand, the principalities and powers we wrestle with may use conflict between the nations just to keep us so angry at one another that the idea of loving our neighbors is unbearable.

One of the interesting finds among the ruins of Mari are the *humûsum*, commemorative funerary monuments for individuals of high status. Ancestors appear to have been venerated at Mari, a practice we find among Amorite societies elsewhere in Mesopotamia. We'll get to the significance of that in a bit.

The Amorites, like the later Israelites, were subdivided into tribes. By the time of Abraham, the Amorites were divided into two main groups— the Binū Yamina, whom we mentioned in the previous section, and the Binū Sim'al ("sons of the left hand," or northerners).

Binū Yamina is the same name Jacob gave his youngest son, Benjamin, although there's no relation between the Amorite tribal group and the tribe of Israel other than language—not that some scholars haven't tried to find a link. (It's also the origin of the name of the country of Yemen. It's as far to the "right" as you can go in Arabia.) The division between the groups seems based on an ancient agreement over pasturing rights: The Bensimalites took their herds to the Khabur River triangle in what is today northern Syria, while the Benjaminites pastured their flocks in the territories of Yamḫad (Aleppo), Qatna, near modern-day Homs, and Amurru, the mountains of northern Lebanon to the southeast of Ugarit.

King Zimrī-Līm of Mari was a Bensimalite. During his reign, the

Benjaminites were divided into five tribes: the Yahrurû, the Yahrihû (who gave their name to Jericho), the Amnanû, the Rabbû, and the Uprapû. To be truthful, Amorite tribal names aren't especially relevant to our discussion—except for one that we'll get to shortly.

It appears that sometime around the end of the third millennium B.C., the Sumerian dynasty of Ur collapsed under the weight of intrusions by Amorite tribes and invasions from Elam. The century beginning about 2000 B.C. is a haze, a period of history that we'll probably never decipher. When the fog lifts around 1900 B.C., we find Amorites ruling nearly all the power centers in Mesopotamia and the Levant. They'd spread out from their traditional base on the steppes of northern Syria and Iraq to as far as the Jordan River in the southwest, southern Turkey in the north, and as far southeast as an insignificant village on the Euphrates that would soon play a major role in world history—some of which hasn't happened yet.

Because of their association with Jebel Bishri and the steppes of Syria, west of Sumer, the Sumerian word for Amorite, MAR.TU, became a synonym for the compass point west, just like the Hebrew link between Mount Zaphon and *tsaphon*, north. However, there was another area occupied by Amorites. Northeast of modern Baghdad, around the Diyala River valley, an Amorite tribe called the Tidnim, Tidnum, or Tidanum was a major source of trouble for the Sumerians at least as far back as 2800 B.C., when the first mention of a "chief" or "overseer" of the tribe is found at Ur. (However, one of the peaks of Jebel Bishri was called Jebel Diddi, or Mount Diddi, which may come from *Didānum*—same name, different transliteration—so the tribe of that name may have history there as well.)[25]

The Tidnum of northeast Mesopotamia were so troublesome that the last kings of Ur built a wall, possibly more than a hundred miles long, from the Euphrates across the Tigris to a site along the Diyala. The project was called the *bad mar-du murīq-tidnim*, or, "Amorite wall which keeps the Tidnum at bay."

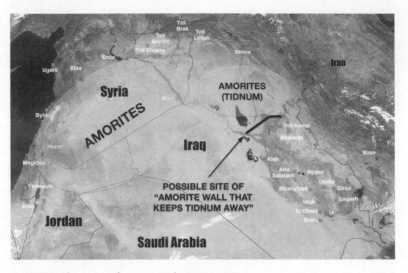

Areas under Amorite control ca. 2000 B.C.

It may have been more of a border fence than a Great Wall of China-type structure. Scholars point out that the only mentions of the wall are the names of the fourth and fifth years of king Šu-sin's reign, and that there are no mentions of it at all among the piles of tablets that recorded the king's business in Ur. However, what Šu-sin *has* done is establish the link between the Amorites and the Tidanum. And that's important.

Whoever the Tidanum were, they were clearly honored—actually, venerated—by their descendants. Based on texts found across the ancient Near East, the Tidanum were the ancestors of the ruling houses of the old Assyrian kingdom of Šamši-Adad, the old Babylonian empire of Hammurabi the Great, and the kings who ruled Ugarit down to about 1200 B.C. Two of the kings in the line of Hammurabi include a variant of the tribal name as the theophoric element (Ammi-ditana and Samsu-ditana), indicating that the Tidanum, if not gods, were at least considered god-like by their descendants.

The name Tidanum may come from an Akkadian word meaning "aurochs," the wild ancestor of today's domesticated cattle. They were bigger and more athletic, truly impressive animals, and probably downright intimidating in an age when the average man stood about 5'6". That was undoubtedly the point.

The name Tidanum and its variants—Tidinu, Titinu, Tidnim, Dida-num, Ditanu, Datnim, Datnam—had a strong military connotation in ancient Mesopotamia. A text from Ebla refers to an official named Tidinu, chief of the mercenaries in the court of Ibrium, the powerful vizier of king Irkab-damu in the twenty-fourth century B.C. It appears that some Amorites served Sumerian kings during the late third millennium B.C. as soldiers, and they might even have been part of the royal bodyguard in Ur for kings Shulgi and Šu-shin.

But the Tidanum were also depicted as enemies of Sumer alongside Anshan (northwest Iran) by king Shulgi, who ruled Ur for nearly half a century sometime around 2100 B.C. His son, Šu-sin, marked one of the years of his reign as the year in which he defeated a coalition of rebellious Tidnum and Ya'madium, another Amorite tribe that lived in eastern Mesopotamia.

And when Ur finally fell around 2004 B.C., the "Tidnumites" were one of the groups blamed in a *Lamentation Over the Destruction of Sumer and Ur*:

> In Ur, no one took charge of food, no one took charge of water,
> Who was (formerly) in charge of food, stood away from the
>   food, pays no heed to it,
> Who was (formerly) in charge of water, stood away from the
>   water, pays no heed to it,
> Below, the Elamites are in charge, slaughter follows in their
>   wake,
> Above, the Halma-people, the "men of the mountains," took
>   captives,
> The Tidnumites daily fastened the mace to their loins.[26]

Given this evidence and a reference by Gudea of Lagash, another city in southern Mesopotamia, that he imported blocks of stone for statues from *kur mardu*, the mountain of the Amorites, it's a good bet that there were two centers of Amorite power at the start of the second millennium B.C., one around Jebel Bishri in Syria and the other around the Diyala valley and Jebel Hamrin in Iraq.

After the fall of Ur, things changed quickly. Scholars still debate whether the change was by conquest or gradual assimilation, but within a hundred years or so Amorites had taken control of every major political entity from Canaan to southern Mesopotamia.

A hundred years after that, by the early eighteenth century B.C., an Amorite city-state that had been a small settlement on the Euphrates since about 2300 B.C. began an incredible rise to become the dominant power in Mesopotamia. It became so influential that people from the time of Isaac and Jacob until the present day have called the region Babylonia instead of Sumer.

Take a moment and think back to Genesis 15:16: "The iniquity of the Amorites is not yet complete."

Now realize that Babylon, a byword for sin, depravity, and occult wickedness throughout the Bible—even in prophecy that is yet to be fulfilled—was founded by the Amorites.

There is no ethnicity called "Babylonian." That's like saying this author is ethnically Chicagoan (Chicagoite?) just because he was born there. Babylon was just the name of a small village led by an ambitious Amorite chief named Sumu-Abum. He and his next few successors didn't even bother calling themselves "king of Babylon," an indication of how unimportant it was at first. It was more than a hundred years before Hammurabi made Babylon a regional power.

But even then, the spiritual significance of Eridu was still supreme: Hammurabi was crowned *there*, the site of Nimrod's Babel, instead of in his home city Babylon, thus linking Enki, god of the *abzu*, to the Amorites.

As with all cities in Mesopotamia, the village of Babylon had a patron god. Marduk was a second-rate member of the pantheon until the Amorites elevated Babylon to prominence during the eighteenth century B.C. By the second half of the second millennium B.C., Marduk had assumed the top position in the Mesopotamian pantheon, replacing Enlil as king of the gods.

Marduk's early character is obscure. As Babylon grew in importance, the god assumed traits of other members of the pantheon, especially Enlil and Ea (the Akkadian name for Enki). Marduk was considered the god of water, vegetation, judgment, and magic. He was the son of Ea and the goddess Damkina, and his own consort was Sarpanit, a mother goddess sometimes associated with the planet Venus, and not surprisingly, the goddess Ishtar/Inanna.

By the time of the Exodus, Babylon had been the dominant power in Mesopotamia for nearly three hundred years. It proudly carried on the tradition of arcane pre-flood knowledge brought to mankind by the apkallu, who, as fish-men from the *abzu*, were presumably able to survive the Flood, or by Gilgamesh, who is mentioned on a cylinder seal as "master of the apkallu."

Of course, the Hebrews under Moses knew that the knowledge of the apkallu/Watchers was forbidden, secrets that humans weren't meant to know. It was as much the reason for the imprisonment of the Watchers as their sexual sin of mating with human women. And this knowledge was proudly celebrated by the Amorite kings and priests of Babylon.

Bringing this back around to the conquest of Canaan: That's why the kingdoms of Sihon of Heshbon and Og of Bashan were *kherem*. And that's why Moses went out of his way to show his readers the connection between the Amorites of Babylon and the Amorites of the Promised Land.

How did he do that? Have you ever wondered why Moses bothered to include this little detail in his account of the victory over Og?

> So we took the land at that time out of the hand of the two kings
> of the Amorites who were beyond the Jordan, from the Valley of
> the Arnon to Mount Hermon (the Sidonians call Hermon Sirion,
> while the Amorites call it Senir), all the cities of the tableland and
> all Gilead and all Bashan, as far as Salecah and Edrei, cities of the
> kingdom of Og in Bashan. (For only Og the king of Bashan was

left of the remnant of the Rephaim. **Behold, his bed was a bed
of iron. Is it not in Rabbah of the Ammonites? Nine cubits was
its length, and four cubits its breadth, according to the common
cubit.**) (Deuteronomy 3:8–11, ESV, emphasis added)

Nine cubits by four cubits is thirteen-and-a-half feet by six feet! So Og
was a giant, right? Well… No, not necessarily.

Yes, the Rephaim were linked to the Anakim, the descendants of the
Nephilim. And tradition holds that Og was a really big dude. But that
wasn't the point here. Moses was writing to an audience that was familiar
with the infamous occult practices of Babylon.

Every year at the first new moon after the spring equinox, Babylon
held a new year festival called the *akitu*. It was a twelve-day celebration of
the cycle of regeneration, the beginning of a new planting season, and it
included a commemoration of Marduk's victory over Tiamat. The entire
celebration, from Yahweh's perspective, was a long ritual for "new gods
that had come recently" involving all manner of licentious behavior.

The highlight of the festival was the Divine Union or Sacred Marriage,
where Marduk and his consort, Sarpanit, retired to the cult bed inside the
Etemenanki, the House of the Foundation of Heaven and Earth, the great
ziggurat of Babylon. Although scholars still debate whether the Sacred
Marriage was actually performed by the king and a priestess, it didn't
matter to Yahweh. The idea that a bountiful harvest in the coming year
depended on celebrating Marduk's sacred roll in the sack was abhorrent.
(And remember, it's possible that this new year festival was celebrated at
Jericho just before the Israelites began marching around the walls.)

Now, here's the key point: Guess how big Marduk's bed was?

"…nine cubits [its long] side, four cubits [its] front, the bed; the
throne in front of the bed."[27]

Precisely the same dimensions as the bed of Og. *That* is why Moses
included that curious detail! It wasn't a reference to Og's height; Moses

was making sure his readers understood that the Amorite king Og, like the Amorite kings of Babylon, was carrying on pre-Flood occult traditions brought to earth by the Watchers.

And that's why Og and his Amorite ally Sihon were *kherem*. It was, and still is today, a spiritual battle.

But that's not all. Oh, no—there's a lot more we can add to the legacy of the Amorites.

———————

Let's bring the timeline into focus: The Bible tells us that Abraham arrived in Canaan 430 years before the Exodus. With the Exodus at 1446 B.C., that puts Abraham in Canaan in 1876 B.C., just as the fog over the political situation in Mesopotamia lifted with Amorites in control.

So let's review:

Subu-Abum founds kingdom of Babylon—1894 B.C.
Abraham arrives in Canaan—1876 B.C.
Isaac born to Sarah—1851 B.C.
Isaac marries Rebekah—1811 B.C.
Šamši-Adad conquers Assyria—c. 1809 B.C.
Hammurabi crowned king of Babylon at Eridu—1792 B.C.
Jacob and Esau born—1791 B.C.
Abraham dies—1776 B.C.
Zimrī-Līm takes throne of Mari—1775 B.C.
Hammurabi dies—1750 B.C.
Hyksos take over Lower Egypt—c. 1750 B.C.
Jacob arrives in Egypt—1661 B.C.
Ahmose drives Hyksos out of Egypt—c. 1550 B.C.
Kassites conquer Babylon—1505 B.C.
Moses leads the Exodus—1446 B.C.
Joshua leads the Conquest—1406 B.C.

Interesting, isn't it, that the Amorite domination of Mesopotamia began just as God called Abraham and directed him to Canaan? And that Babylon reached the peak of its power with the ascension of Hammurabi the Great just about the time Jacob and Esau were born?

At the same time, a Semitic-speaking, Ba`al-worshiping state emerged to take control of northern Egypt just before the arrival of the house of Jacob. Wouldn't it be, you know, *coincidental* if the Hyksos rulers of Lower Egypt were Amorites, too?

Well, yes, it would—if we believed in coincidences. And as it happens, scholars do, in fact, believe the Hyksos were Amorites.

The best-known of the Hyksos kings, Khyan, is attested from inscriptions found as far from Egypt as Cyprus and modern-day Baghdad (probably originally in Babylon). An Amorite king with the same name, spelled Hayanu, is listed in a genealogy as a distant ancestor of Šamši-Adad and the royal house of the old Assyrian kingdom.

Scholars have also noticed strong similarities in the burial practices of the Hyksos and various Amorite kingdoms, especially the practice of sacrificing donkeys for burial with important people and under the doorways of new buildings.

So yes, it's strangely coincidental, if you're a believer in coincidence theories. Otherwise, it seems an unseen hand or hands moved the Amorites into position in Egypt and Canaan just before the Israelites arrived—almost as if they'd been placed there to wait for God's chosen people.

It's also noteworthy that while the Anakim were confirmed in Canaan by extrabiblical sources from Egypt, the Anakim haven't been found anywhere else in the ancient Near East.

But you know by now we're not coincidence theorists. Let's pull these historical threads a little harder Why did God link the timing of Israel's return to the iniquity of the Amorites?

We've established that the Amorites founded Babylon. This is documented mainstream history, unquestioned by secular scholars. The Amorites are also linked by historic records to a tribe called the Tidnum/

Tidanum, which seemed to have a strong military reputation, troublesome enough that the last Sumerian kings of Mesopotamia built a very long wall in a futile attempt to keep them away. That is also solidly documented mainstream history.

The Bible links the Amorites to the Anakim (Deuteronomy 2:10–25). The Transjordan campaign was aimed at two Amorite kings, Sihon and Og, who were both remembered as giants. Og gets most of the attention from Bible students today because he was called the last of the Rephaim, but the Bible records a song about the defeat of Sihon that was apparently well-known in Old Testament days:

> For Heshbon was the city of Sihon the king of the Amorites, who had fought against the former king of Moab and taken all his land out of his hand, as far as the Arnon.
> Therefore the ballad singers say,
> "Come to Heshbon, let it be built;
> let the city of Sihon be established.
> For fire came out from Heshbon,
> flame from the city of Sihon.
> It devoured Ar of Moab,
> and swallowed the heights of the Arnon.
> Woe to you, O Moab! You are undone, O people of Chemosh!"
> He has made his sons fugitives,
> and his daughters captives,
> to an Amorite king, Sihon.
> (Numbers 21:26–29, ESV)

Note the acknowledgement that Chemosh was the national god of Moab. Obviously, Chemosh wasn't up to the test of whatever god was backing Sihon and his Amorites. Maybe Sihon and his northern neighbor Og were drawing on dark powers—even darker than Chemosh, who went in for child sacrifice on occasion (2 Kings 3:27).

Remember that the Amorite king Og was called the last of the Rephaim

(Deuteronomy 3:11). Texts found throughout Mesopotamia identify the Tidnum/Tidanum as a tribe of the Amorites. In Ugarit, the spelling is usually Ditanu or Didanu. And those texts are usually in the context of venerated dead ancestors, especially the honored royal dead.

The city-state of Ugarit, usually described as Canaanite by Bible teachers, was more accurately an Amorite kingdom. The tablets found by archaeologists there have provided a wealth of knowledge about ancient Hebrew and the history of the period around the time of the Exodus. And some of the connections between secular history, the Bible, and the supernatural realm are absolutely fascinating.

For example, a funerary text identified as KTU 1.161, or RS 34.126 (designations that identify the tablet in question), plainly connects the Amorite tribe of Didanu (Ditanu/Tidnum/Tidanum) with the biblical Rephaim—in a ritual to summon them from the dead!

> "Sacrifice of the Shades" liturgy:
> **You are summoned, O Rephaim of the earth,**
> **You are invoked, O council of the Didanu!**
> *Ulkn*, the *Raphi'*, is summoned,
> *Trmn*, the *Raphi'*, is summoned,
> *Sdn-w-rdn* is summoned,
> *Tr 'llmn* is summoned,
> **the Rephaim of old are summoned!**
> **You are summoned, O Rephaim of the earth,**
> **You are invoked, O council of the Didanu!**[28]
> (Emphasis added)

Yeah, I know! *What?!*

Scholars who look at this text from a secular perspective tend to view it as an academic curiosity, a window into the psychology of people who lived 3,500 years ago. But as Christians, filtering this through the lens of truth, we get a whole different picture.

In this ritual, the Rephaim, which included a council of the Didanu, were invoked to accompany the recently deceased king of Ugarit, Niqmaddu III, to the underworld. They were also there to bless the new king, Ammurapi III, who—although he probably didn't know it—was the last king of Ugarit. His kingdom was about to be overrun by the so-called Sea Peoples sometime around 1200 B.C.

The Rephaim and the council of the Didanu were apparently summoned to impart to the king the power to overcome death, and to make the living king one of the *rpum*—the Rephaim. Assuming this ritual wasn't an invention for Ammurapi, and evidence from Babylon suggests it wasn't, it appears that the coronation rites of the Amorite kings of Ugarit (and maybe other Amorite kingdoms) summoned the king's dead ancestors, *whom they identified as the Rephaim.*

Imagine a ritual like that in front of the White House on Inauguration Day!

Now, buckle up, because we're going to work through some challenging stuff in the rest of this chapter. Most of what follows comes from a paper published in 1999 by scholar Amar Annus of the University of Tartu, Estonia,[29] whose research into the Mesopotamian origins for some of the weirder themes in the Old Testament, like the Watchers and their sin, is truly groundbreaking.

For starters, Annus concluded that the West Semitic root for the word Rephaim, *mrp'*, appears to be the origin of the Greek word *merops*. The word can have a similar meaning in both languages, "healer" or "healing."

Kos, an island in the southeast Aegean Sea off the coast of Turkey, was formerly called Meropis, after one Merops, the mythical first king of the island. Merops was thought to be an autochthon, an original inhabitant of the land, one who sprang from the rocks and trees as opposed to a foreigner who settled in it. His people, then, were the Meropes. The key point is that Merops, Meropis, and the Meropes all derive from the Semitic root *mrp'*.

We also find that root as the basis of the phrase *meropes anthropoi*. That phrase was used by the Greek poet Hesiod in his famous poem *Works and Days* to describe the men who lived in a long-ago Golden Age.

First of all the deathless gods who dwell on Olympus made a golden race of mortal men who lived in the time of Cronos when he was reigning in heaven. And they lived like gods without sorrow of heart, remote and free from toil and grief: miserable age rested not on them; but with legs and arms never failing they made merry with feasting beyond the reach of all evils. When they died, it was as though they were overcome with sleep, and they had all good things; for the fruitful earth unforced bare them fruit abundantly and without stint. They dwelt in ease and peace upon their lands with many good things, rich in flocks and loved by the blessed gods. (Hesiod, *Works and Days*)

Who were these *meropes anthropoi*? Homer named a few of them in *Odysseus*: Theseus, who killed the Minotaur on Crete; Aegeus, the mythical founder of Athens; Polyphemus, the cannibalistic giant son of Poseidon, one of the Cyclopes; Caneus, a nigh invulnerable warrior, transformed from a woman into a man by Poseidon; Dryas, leader of a tribe that fought a long war with the Centaurs; and so on.

Significantly, Hesiod mentions that the *meropes anthropoi* became, upon death, *daimones*, although he viewed them more favorably than Jews and Christians do demons:

But after earth had covered this generation—they are called pure spirits dwelling on the earth, and are kindly, delivering from harm, and guardians of mortal men; for they roam everywhere over the earth, clothed in mist and keep watch on judgements and cruel deeds, givers of wealth. (Hesiod, *Works and Days*)

The Book of Enoch offers a slightly different explanation for the origin of demons:

And now, **the giants, who are produced from the spirits and flesh, shall be called evil spirits upon the earth, and on the earth**

shall be their dwelling. Evil spirits have proceeded from their bodies; because they are born from men and from the holy Watchers is their beginning and primal origin; they shall be evil spirits on earth, and evil spirits shall they be called. [As for the spirits of heaven, in heaven shall be their dwelling, but as for the spirits of the earth which were born upon the earth, on the earth shall be their dwelling.] And the **spirits of the giants afflict, oppress, destroy, attack, do battle, and work destruction on the earth, and cause trouble:** they take no food, but nevertheless hunger and thirst, and cause offences. And these **spirits shall rise up against the children of men and against the women, because they have proceeded from them.** (1 Enoch 15:8–12, R. H. Charles translation, emphasis added*)*

Thus, between Hesiod and Enoch we can connect the *meropes anthropoi*, the men of the Golden Age, to the Nephilim, children of the fallen Watchers. Both lived during a pre-Flood age, and both, upon death, became wandering spirits called demons. It's just that the Greek view of *daimones* was more favorable than the Jewish (or Mesopotamian, for that matter) understanding of demons.

Needless to say, that's another PSYOP by the Enemy.

Kronos, Saturn to the Romans, was king of a race of gods called the Titans, who reigned supreme after Kronos deposed his father, Uranus (with extreme prejudice—Kronos castrated him with a scythe). The golden race of men created by the Titans was the only one that lived during the reign of Kronos. Told that he would be deposed in turn by his children, Kronos tried to preserve his kingship by eating his kids as soon as they were born. Zeus was spared that fate by his mother, Rhea, who gave Kronos—obviously not a picky eater—a boulder wrapped in a blanket instead. When Zeus was grown, he freed his siblings and led a war to depose the old tyrant. The Titans were defeated and imprisoned in Tartarus.

So through the link between the Semitic root *mrp'* and the Greek word *meropes*, we have a connection between the Nephilim and the heroic men of the Golden Age of Kronos, the *meropes anthropoi*, "the mighty men who were of old, the men of renown."

Now, back to the Bible: Og, last "of the remnant of the Rephaim," ruled the land of Bashan, a territory that included Mount Hermon, the place where the rebellious Watchers descended. In Deuteronomy 1:3, Joshua 12:4–5, and Joshua 13:12, we're told specifically that Og "lived at Ashtaroth and at Edrei and ruled over Mount Hermon." Edrei was the site of the battle between Israel and the forces of Og. And one of the Ugaritic texts, KTU 1.108, confirms the link between Og and the Rephaim as denizens of the netherworld.

> May **Rapiu, king of eternity**, drink wine, may he drink, the powerful and noble god, **the one who rules in Athtarat** [Ashtaroth], **the god who reigns in Edrei**, who sings and plays on the lyre. (KTU 1.108, emphasis added)

In other words, the Amorites of Ugarit believed that a god named Rapiu, a singular form of the word *rpum* (Rephaim), ruled *exactly the same territory* as Og, king of Bashan. And since Rapiu, the king of eternity, was linked to the Rephaim, the honored ancestral dead, Og's kingdom around Mount Hermon was essentially the gateway to the underworld.

Here's another interesting data point: In Ugaritic, Bashan, which Ugaritians pronounced with a "th" instead of an "sh." meant "place of the serpent"—a callback to the divine rebel, the *nachash*, of Genesis 3. Remember from Isaiah 14, the nachash was cast down to Sheol where the dead kings of the nations reside. Did that happen at Bashan?

Canaanite myth offers another link between Og and the Rephaim: Danel (the Ugaritic equivalent of the Hebrew name Daniel), the hero of a Canaanite myth called *The Legend of Aqhat*, is described in the story as a *mt rpi*. According to Amar Annus, *mt rpi*, which means "man of Rephaim." is a linguistic match for *meropes anthropoi*. That specifically links the golden race men from the age of Kronos—i.e., the Nephilim—to the Rephaim, and thus to the council of the Didanu.

But get this: Danel is also called *mt hrnmy*, which means "man of Hermon."

Yeah. *That* Hermon.

So now we can link the biblical Rephaim, the mythical *meropes anthropoi* of the Golden Age of Kronos, the Nephilim, the Watchers of Genesis 6, and the mysterious council of the Didanu—which, remember, was probably the name of an ancient tribe of Amorites claimed as the ancestors of the kings of Ugarit, Assyria, and Babylon.

This is a good time to point out that the ill-fated Ammurapi III of Ugarit mentioned above shared a name with the most famous king of the old Babylonian empire, Hammurabi. Scholars typically translate their names, *ammu rapi*, as "my kinsman is a healer." This draws on the possible meaning "healer" of the Semitic root *rpi*.

Although this author is not a scholar of ancient Semitic languages, in the context of what we've just read, a more accurate rendering of Ammurapi/Hammurabi might be "my kinsman is a *rapha*"—one of the Rephaim.

File a mental bookmark there. We'll come back to it.

An Akkadian lexical list called *Malku = šarru* equates the words Amorite, Tidnu, and Sutu, the latter word being the name of yet another Amorite tribe, the Suteans. It appears that by the time of the Exodus, "Sutean" was a general term for nomadic shepherds of the Syro-Arabian desert.

Not all scholars agree that the Suteans were technically Amorites, but the Akkadian scribes thought so, and furthermore linked them specifically to the Tidnu (Didanu). The tribe apparently wandered western Mesopotamia between the Jordan valley and the Euphrates, a thorn in everybody's side. During the early Old Babylonian period, the Suteans were known to the Near Eastern world as warriors and robbers. However, by the later Old Babylonian period Suteans were connected with royal military service, mainly for the city-state of Sippar, the cult city of the sun god Utu (Sumerian)/Šamaš (Akkadian), located on the east bank of the Euphrates about thirty-five miles north of Babylon.

Once again, the Egyptians confirm the existence of a group of people in a time and place that's relevant to our study. Execration texts from about the nineteenth or eighteenth century B.C. mention several princes of the Suteans, two of which have names that should be familiar to readers of the Bible.

> The Ruler of Shutu, Ayyabum, and all the retainers who are with him; the Ruler of Shutu, Kushar, and all the retainers who are with him; the Ruler of Shutu, Zabulanu, and all the retainers who are with him.[30]

Ayyabum is an archaic form of Job. Interestingly, this text is from about the right time and it places Ayyabum in the right location to be the Job of the Bible (although we're not making that claim, just demonstrating that the Bible's context is historically accurate). Zabulanu is obviously the Semitic name Zebulon, although it's not likely he was the son of Jacob. Kushar, some scholars believe, may be the same name as Cushan in Habakkuk 3:7, where it's linked to the land of Midian, also in the Transjordan.

The Shutu, Egyptian *šwtw*, are the Suteans, and the texts indicate a location in the Transjordan, and most likely in the *northern* Transjordan—exactly where the Israelites encountered Sihon of Heshbon and Og of Bashan.

Scholars argue the point, but there is evidence to suggest that Shutu is a variant spelling of Shasu, the extremely tall prisoners we met earlier on the stele depicting the Battle of Kadesh. Others suggest that the Shutu are the progenitors of the Moabites, because the Egyptian texts and the Bible locate the Shutu and Moab in the same place.

Now here's where things get a little speculative: Scholars have suggested that the name of the Sutean tribe comes from the name Seth:

Sutī'ū was the usual term in Akkadian for Western Semitic shepherd tribes in Mesopotamia. This is apparently the transcription of Amorite *Šetī'u*, which means "descendants of Šutu/Šitu." The name Suteans most probably derives from the same root as the biblical name Seth (*Šēt*), son of Adam.[31]

If that's correct, then the Suteans are mentioned in the messianic prophecy of Balaam, son of Beor, which he delivered, appropriately enough, while Israel was in Moab.

I see him, but not now;
I behold him, but not near:
a star shall come out of Jacob,
and a scepter shall rise out of Israel;
it shall crush the forehead of Moab
and break down all the sons of Sheth.
Edom shall be dispossessed;
Seir also, his enemies, shall be dispossessed.
Israel is doing valiantly.
(Numbers 24:17–18, ESV, emphasis added)

So this is perhaps a prophecy that the sons of Sheth, the Shutu/Suteans, identified by Akkadian scribes as the Amorite tribe Tidnu (Didanu), will be destroyed by the Messiah.

Amar Annus thinks the sons of Sheth, the Suteans, were so called because they traced their lineage back to Seth, son of Adam. Well, yes, but since Noah was descended from Seth, so does everybody else, including you and me.

This is speculation, but considering that the Shasu/Shutu/Suteans were in contact with Egypt, where Amorite rulers who worshiped the god Set had been in power for a couple hundred years just prior to the Exodus, let's consider another idea: Since Set, identified with Ba`al in the Hyksos pantheon, is also spelled Seth (and Setesh, Sutekh, Setekh, or Suty) and was a god of the desert, storms, disorder, and violence, the warlike, disruptive Suteans may have been "sons of (the god) Seth" rather than "sons of (the patriarch) Seth."

Indeed, some English translations of Numbers 24:17 render *bene sheth* as "sons of disorder," which is appropriate for desert-dwelling followers of the Egyptian god of chaos.

Significantly, one of the Akkadian exorcism texts called *Maqlû* ("burning") names Suteans as practitioners of evil witchcraft. So now our links extend further into the realm of the supernatural. Let's run all of this down and try to make sense out of it.

The kings of the Amorites in Ugarit, Assyria, and Babylon were descended from an ancestral tribe also called the Didanu/Tidanum. That tribe, also called the Tidnum, was equated with a nomadic Amorite tribe called Suteans in an Akkadian lexicon. The Suteans appear to be the Shasu/Shutu of Egyptian texts and the "sons of Sheth" in Numbers 21. They were located in the Transjordan by the Bible and Egyptian texts.

The Bible links the Amorites to the Rephaim and the Anakim, descendants of the Nephilim. A ritual text from Ugarit connects Amorite kings to the Rephaim and the council of the Didanu, apparently spirits of the tribe's honored dead ancestors, who were summoned to convey blessings on new kings.

Through Hesiod, we can link those spirits, the council of the Didanu and the Rephaim, to the mythical *meropes anthropoi*, mighty men of a pre-Flood golden age. The spirits of those men became helpful *daimones* upon death, while the spirits of the monstrous pre-Flood Nephilim became evil demons upon death.

Clear?

Now, let's go back to that mental bookmark. Since you're perceptive, you've probably already figured out where this is leading. But to put this on the record, we will now lay this out in black and white: The name of an ancestor of several Amorite royal houses, Dedan, whose descendants were called the Didanu, Tidanum, and variations thereof, is the name from which the Greeks derived the word *titanes*—from which we get the name of the Titans.

Dedan is a name attested in the Bible. Dedan and Sheba are locations in western Arabia mentioned several times by the prophet Ezekiel (about which more later). It's also the name of one of the leaders of Korah's rebellion against Moses, Dathan (see Numbers 16).

One of the nephews of Nimrod was named Dedan, maybe not coincidentally. Could his name have been in honor of the Titans, the old gods who descended at Mount Hermon in the dim, distant past? Given the later links between the Amorites and the Rephaim, and the pre-Flood knowledge brought back into the world by the Amorite kingdom of Babylon, this speculation isn't exactly coming out of thin air.

While we'll never identify for sure *the* Dedan whose name became synonymous with the old gods, or why the Amorites appear to have carried the belief that they were their heirs, we can document the connection between the Watchers of the Bible and the Titans of Greek myth.

When the Olympians defeated the Titans, Zeus banished them to Tartarus, a place of torment for the wicked as far below Hades as the earth is below heaven. That just happens to be the current address of the Watchers who landed at Mount Hermon.

> For if God did not spare angels when they sinned, but **cast them into hell** and committed them to chains of gloomy darkness to be kept until the judgment. (2 Peter 2:4, ESV, emphasis added)

The Greek word translated "cast them into hell" is the verb ταρταρόω, *tartaroo*, which literally means "to thrust down to Tartarus." This is the only use of that word in the New Testament. Hades, meanwhile, is mentioned nearly a dozen times, including twice by Jesus. That distinguishes Hades from Tartarus, which was apparently reserved as a special place of punishment for angels who sinned. And the only explicit example of angels sinning in the Bible is in Genesis 6:1–4, which is confirmed by the passages in 2 Peter and by Jude, who clearly linked the punishment of the angels to a sexual sin.

Further, the Jewish scholars who translated the Hebrew Scriptures into Greek two hundred years or so before Jesus' birth into the text called the Septuagint understood the link between the Rephaim and the pagan gods. In 2 Samuel 5, the scholars translated the site of David's battle against the Philistines, *emeq rapha*, as Valley of the Titans.

So. The Titans of the Greek myths were the "angels who did not stay within their own position of authority," the Watchers of Genesis 6. They are bound in Tartarus, kept "in eternal chains under gloomy darkness until the judgment of the great day." Their children, the Nephilim, whose spirits are the demons that plague the earth to this day, were the shades of Sheol, the Rephaim, who were summoned in rituals by Amorite kings who believed they were their honored dead ancestors.

Seriously.

Does the phrase "the iniquity of the Amorites" begin to make more sense?

SUMMING UP

We covered a lot of ground in this chapter. The seat of Ba`al's power, the site of his palace, was known to the ancient world as Mount Zaphon in modern-day Turkey. It was also the location of the holy mountain of the Hurrian and Hittite storm-gods, Teššub and Tarhunt, and it was holy to Zeus and Jupiter, too. No wonder, since they're all the same god.

Ba`al was the main antagonist of the Hebrews throughout the Old Testament, and was still in the land during the time of Jesus. More on that later.

The conquest of Canaan was, like the crossing of the Red Sea, a supernatural conflict. The Israelites were ordered to destroy the kingdoms of the Amorite kings Sihon and Og in the Transjordan, followed by a divide-and-conquer strategy aimed at the Amorites and Anakim of Canaan.

The attack on Jericho, a cult center of the Amorite moon-god Yarikh, was specifically timed to coincide with the first full moon of the new year. Joshua's Long Day, the battle in the Valley of Aijalon against the Amorite coalition led by the king of Jerusalem, may have been a further insult to the moon and sun gods; Yahweh stopped them from keeping their appointed times in the sky until the Amorite armies were utterly destroyed—*kherem.*

We then spent a good deal of the chapter examining why the Exodus and Conquest happened when they did, when the iniquity of the Amorites was apparently complete in the eyes of Yahweh. The sin of the Amorites has multiple layers: They were responsible for founding Babylon and its occult wickedness, which is still infamous today nearly four thousand years later. Moses specifically linked the occult bed inside the temple of Marduk in Babylon to the bed of the Amorite king Og to make that point clear.

The Amorite kings also apparently considered themselves the descendants and heirs of the Titans (the Watchers) and their children, the Nephilim. The Amorites summoned their venerated dead ancestors, the Rephaim and the council of the Didanu, in rituals to give their kings power in the natural realm.

The Amorites occupied the Holy Land and stood in opposition to the Israelites specifically because they were the agents of spiritual forces that wanted to keep Yahweh and His people away from His holy mountain, Zion.

———◆———

One final nugget for this chapter: Moriah, the name of the mountain where Abraham was tested by Yahweh more than four hundred years before the Conquest, may have meant "land of the Amorites." While the consensus among Christian scholars seems to be that Moriah means "chosen by YHWH," some believe the first syllable of the word was lost through aphesis, where an unstressed vowel at the beginning of a word is dropped over time—like "round" from "around."

Regardless, when you peel away the layers, you find the history in the Bible all about the mountain—*His* mountain. And contrary to the belief of secular historians, the Amorites did not fade into history after Israel took Canaan. That's another bit of supernatural disinfo. The Amorites—and the spirits behind them—are still out there.

6

CARMEL

THE NEXT MOUNTAIN in our historical survey is one that was almost over-looked in the research for this book. In fact, this author completely missed it in the first public presentation that led to writing this book. Thankfully, this author's wife is a lot smarter than he is and grasped the significance of this well-known Bible story and convinced him to dig into it.

We're going to skip ahead about six hundred years. A lot of good stuff will be passed over, like Ezekiel's wheel and Daniel's experience in Babylon. A good case can be made that the writing on the wall during the feast of Belshazzar—*Mene, mene, tekel, upharsin*—was divine judgment by Yahweh on the chief god of Babylon, Marduk. That must wait for a future book, God willing.

The inspiration for this book was a look at divine mountains and the supernatural battles that took place around them, so some history in between those key points is being left for future projects. We beg your indulgence as we rocket through the next six centuries in just a few pages.

After Israel took possession of the land, which a careful reading of the Old Testament informs you wasn't exactly complete, the Israelites had some rest for a while. Not long, though; the supernatural enemies of Yahweh have nothing to gain by giving up.

As we mentioned earlier, the Semitic storm-god, Ba`al-Hadad, began moving into Israel from the north shortly after the Conquest. Apparently being humiliated at the Red Sea and the Jordan River wasn't enough to keep Ba`al from sending his minions back into the land to try to destroy the house of Israel.

An enemy with a familiar name and in a familiar place put the screws to the northern tribes during the time of the prophetess Deborah. Jabin, king of the powerful city-state Hazor, "oppressed the people of Israel cruelly for twenty years" (Judges 4:3).

Skeptics note that Jabin of Hazor was also the name of a king defeated by Joshua during his northern campaign. So they conclude that the biblical account must be wrong. And since, as filmmaker Timothy Mahoney showed in his excellent documentary *Patterns of Evidence*, they're usually looking in the wrong era for evidence, like the destruction layer at Hazor, they declare that the Old Testament is fiction.

Archaeologists have found, however, evidence that Jabin was a dynastic name, used over and over by rulers of Hazor. Three separate inscriptions referring to the king of Hazor have been found and all of them refer to a king named Ibni-Addu. That's the Akkadian form of the West Semitic name Jabin-Hadad. Two of the inscriptions are three or four hundred years earlier than the conquest of Canaan, but the third inscription dates to the reign of Ramesses the Great. That one fits our timeline for the battle led by Deborah and Barak.

Megiddo is a popular place to hold a battle. The Egyptian pharaoh Thutmose III, one of the candidates for the pharaoh of the Exodus, battled a coalition of Amorites and Canaanites around 1457 B.C. The last major battle of the Sinai and Palestine Campaign in World War I was also fought there between the British and the Ottoman Empire. But at some point in the thirteenth century B.C., an army of angels did battle by the waters of Megiddo!

By the way, it is popularly believed that the final battle between the forces of Antichrist and the heavenly host, the battle of Armageddon, will be fought at Megiddo. That happens not to be the case. We'll deal with that later in the book.

Angelic intervention isn't an aspect of the battle that gets discussed much in church, but Deborah, a remarkable woman by any measure, took note of it in her victory song:

> The kings came, they fought;
> then fought the kings of Canaan,
> at Taanach, by the waters of Megiddo;
> they got no spoils of silver.
> **From heaven the stars fought,**
> **from their courses they fought against Sisera.**
> The torrent Kishon swept them away,
> the ancient torrent, the torrent Kishon.
> March on, my soul, with might!
> Then loud beat the horses' hoofs
> with the galloping, galloping of his steeds.
> (Judges 5:19–22, ESV, emphasis added)

Some respected Bible teachers believe God sent a sudden, powerful rainstorm to slow the chariots of Sisera. That's possible; no general worth his salt would bring chariots onto a muddy field, and Sisera seems to have been the power behind the throne of Jabin. And it would be just like Yahweh to defeat followers of the storm-god by sending an unexpected downpour to take away their tactical advantage.

Here's a clue that this may have been the case. Sisera is the name of a Minoan god, *(j)a·sa-sa-ra*, who can be identified as Zeus Kretogenes ("born on Crete").[32] Zeus, the storm-god, is Ba`al. So using rain as a weapon against the army of the storm-god was an unmistakable message in both the physical and supernatural realms that Ba`al had no power on

Yahweh's allotted land, Israel. And the fact that the army of Israel was led by a man named Barak ("Lightning," the weapon of the storm-god) is just too perfect for words.

The text *does* say that Barak and his men literally got help from above. As we mentioned earlier, angels are often called stars in the Old Testament. Obviously, the twinkly lights in the night sky didn't do battle against the Canaanite army; it was divine assistance, the heavenly host—which, frankly, is the only way a force of infantry, even of ten thousand men, routs an army fielding nine hundred iron chariots.

There is another, admittedly speculative, take on the angelic host. The Hebrew word translated "courses," *mecillah*, usually means a highway or causeway. But in 2 Chronicles 9:11, it's used to describe a terrace or stairway that led up to Solomon's Temple. That evokes an image of Jacob's Ladder, a passageway from the heavenly realm to earth, which, in the context of angels fighting against the forces of Sisera (or the supernatural forces behind them), might be appropriate.

Interestingly, in rabbinical thought, the inhabitants of Meroz cursed by the angel of Yahweh in Judges 5:23—there's Jesus again, another Christophany—were extraterrestrials. Because verse 20 refers to the stars fighting from heaven, the rabbis thought Meroz must be a star system with angelic inhabitants who refused to answer Yahweh's call.

For the record, we lean toward Meroz just being a local village of Israelites.

Well, after forty years of peace, the Israelites got fat and happy again. In a pattern repeated throughout the Old Testament, God allowed Israel to fall under the dominion of the Midianites and Amalekites.

It's a bit of a mystery that God chose Gideon out of all Israel to deliver His people. Gideon's father, we learn from the text, was a follower of Ba`al. Before Gideon could lead his followers into battle, he had to do some divinely decreed landscaping.

That night the LORD said to him, "Take your father's bull, and the second bull seven years old, and pull down the altar of Baal that your father has, and cut down the Asherah that is beside it and build an altar to the LORD your God on the top of the stronghold here, with stones laid in due order. Then take the second bull and offer it as a burnt offering with the wood of the Asherah that you shall cut down."

So Gideon took ten men of his servants and did as the LORD had told him. But because he was too afraid of his family and the men of the town to do it by day, he did it by night. (Judges 6:25–27, ESV)

You'll note in the biblical account that the first conversation Gideon had was with the angel of Yahweh. That's yet another Old Testament appearance of Jesus, the Second Power in heaven. Skeptics who deny the doctrine of the Trinity just aren't paying attention.

To his credit, Gideon's father, Joash, refused to turn him over to the men of the town who wanted to stone Gideon for destroying the altar of Ba`al. It was a brilliant reply: "If he is a god, let Ba`al contend for himself."

You know the rest of the story. With the help of Yahweh, Gideon and his three hundred won a smashing victory over the Midianites, Amalekites, and their allies from the East, who may have included the Suteans or remnants of other Amorite tribes.

Yahweh's purpose in arranging the victory this way was to demonstrate His power, not to establish Gideon's reputation as a military tactician. Sadly, the Israelites didn't quite get the message.

Then the men of Israel said to Gideon, "Rule over us, you and your son and your grandson also, for you have saved us from the hand of Midian."

Gideon said to them, "I will not rule over you, and my son will not rule over you; the LORD will rule over you." And Gideon said to them, "Let me make a request of you: every one of you give me the earrings from his spoil." (For they had golden earrings, because they were Ishmaelites.)

And they answered, "We will willingly give them." And they spread a cloak, and every man threw in it the earrings of his spoil. And the weight of the golden earrings that he requested was 1,700 shekels of gold, besides the crescent ornaments and the pendants and the purple garments worn by the kings of Midian, and besides the collars that were around the necks of their camels.

And Gideon made an ephod of it and put it in his city, in Ophrah. And all Israel whored after it there, and it became a snare to Gideon and to his family. (Judges 8:22–27, ESV)

The "draft Gideon" movement was misguided to say the least. Surely, some of them must have realized that no three hundred soldiers, no matter how good their general was, beats a force of 135,000 (per Judges 8:10) without Yahweh behind them. But they put their trust in Gideon instead.

Second, Gideon, frankly, wasn't worthy of the job. The value of the gold at today's rate is about $700,000. That's not exactly a selfless act. (But let's be honest—who among us has the moral strength to turn down a reward that big?) But even worse, Gideon made an ephod from the gold. That's a garment that was only supposed to be worn by Levites, descendants of Aaron. That was a *serious* violation of the Law. So while Gideon said he wouldn't be king, he sure acted like one.

And what little we read about Gideon and his kids bears that out. As soon as Gideon died, Israel turned back to the Baals. Then his half-Canaanite (i.e., half-Amorite) son, Abimelech—whose name means "my father the king"—led an uprising, slaughtered sixty-nine of his seventy half-brothers, and set himself up as a king.

You can just imagine the discussion in the divine council over that development. "Are you *sure* this is part of Your plan?"

———————

We'll advance the calendar past the anointing of Saul and the rise of David. Again, there's a lot of *there* there, but we're moving ahead to the divided kingdom and another iconic confrontation between a true servant of Yahweh and worshipers of the storm-god, Ba`al. And there's a surprise twist to this familiar story that you probably don't see coming.

By the early ninth century B.C., the united monarchy was already split. The northern tribes had elected to follow Jeroboam, who, not surprisingly, set up golden calves at Bethel and Dan. That was a political move; he realized that if his people kept the feasts at Jerusalem, as Yahweh commanded, his throne would be in jeopardy. People under his rule would be traveling into the territory of his rival, Rehoboam of Judah, and subjected to southern propaganda—not to mention the fact that Jeroboam and his ruling clique would miss out on revenue from pilgrims coming to sacrifice at his high places.

The calves, of course, were symbols of Ba`al. To make matters worse, Jeroboam appointed priests from among the people—non-Levites—and performed some of the sacrifices himself. (At least, as far as we know, he never claimed the calves mysteriously jumped out of a fire.)

Anyway, it's depressingly predictable that while the southern kingdom endured its share of bad kings in terms of their loyalty to Yahweh, the northern kingdom didn't produce a single good one during its short history. And the influence of Ba`al, encroaching from the north through Israel's Aramean and Phoenician neighbors, reached a peak during the reign of the most powerful king of the northern kingdom, Ahab.

You probably already know that Ahab married the daughter of the Phoenician king of Tyre, Jezebel. Her father, a king named Ithobaal I, was a priest of Astarte, the Semitic manifestation of Inanna/Ishtar, goddess of sex and war. Ithobaal, whose name means "with Ba`al," was king of Tyre from 878 B.C. to 847 B.C. During his reign, Tyre expanded its influence across most of what is today Lebanon and coastal Syria and part of Cyprus, and planted colonies in northern Africa. A Phoenician author quoted by the first-century Jewish historian Josephus wrote that Ithobaal came to power by murdering the previous king, Phelles, who'd been on the throne for only eight months. (And Phelles had become king by killing his older brother, Astarymus.)

Ithobaal's priestly calling and his method of acquiring power explains a lot about Jezebel. She was nothing if not motivated when it came to promoting the pagan gods of her father's kingdom, and she wasn't shy about using any means necessary to do it.

Ahab was the son of Omri, a former army commander who'd seized power from Zimri, a chariot commander who assassinated the previous king, Elah. Zimri reigned for all of seven days before Omri was proclaimed king by the army. A showdown between factions of the military ended with Zimri burning himself alive inside the king's house in the city of Tirzah. Omri then bought a hill from a man named Shemer and turned it into his royal city, Samaria. (Shemer probably had no idea that his name would live on for nearly three thousand years. A small group of Samaritans still lives in Israel, fewer than a thousand, convinced that Yahweh's holy mountain is Gerizim, not Zion. But I digress.)

Ahab ruled from about 871 B.C. to 852 B.C., give or take a few years. He turned Samaria into a world-class city during his reign, building the famous "ivory house" mentioned in 1 Kings 22:39. Archaeologists in the 1930s found some of the ivory that decorated the walls and furniture of his palace.

We know a lot about this period of history from Assyrian inscriptions discovered by archaeologists. They usually describe overwhelming victories for their kings, which is sometimes not exactly how things played out. Ahab was a member of a coalition of kings in the Levant who confronted Shalmaneser III at the Battle of Qarqar in 853 B.C. While the commemorative monolith claimed a great victory for Assyria, the fact that Ahab, Hadadezer of Damascus, and the rest kept their thrones, and that Shalmaneser was back fighting in Syria in 849 B.C., 848 B.C., and 845 B.C. suggests otherwise.

The Bible tells us that Ahab was the worst of the kings of Israel up to the time of his reign. Considering the bloody history of regime change in the northern kingdom, that's saying a lot. We're told he built a temple for Ba`al in Samaria and erected an Asherah next to it. Now, scholars don't agree on what an Asherah is. The Canaanite mother goddess Asherah was the consort of El (not to be confused with Astarte, who was sometimes the consort of Ba`al), but the Asherim on the high places around Israel and Judah? We don't know what they were. They may have been some type of consecrated pole. Their remains have been found all over the Holy Land, but you can only tell so much from a posthole and rotted wood. All we know for sure is that they involved worship of a pagan god, and Yahweh did not approve.

Still, even as they followed Ahab, Yahweh didn't abandon the people of Israel. Despite the threat from Assyria, there was frequent war between Israel and the Aramean kingdom of Damascus. In 1 Kings 20:23–43, we read an account of a battle between Ahab and Ben-Hadad of Aram, his former ally at the Battle of Qarqar. It's worth mentioning because of the direct involvement of God in the battle.

The Arameans had suffered a humiliating defeat to Ahab the previous year because Ben-Hadad and his officers were too busy getting drunk to prepare for battle. The hand of Yahweh was in this; a prophet told Ahab personally that Yahweh would deliver the much larger Aramean force into the king's hand, and so He did.

The advisors to king Ben-Hadad ("son of Hadad") were convinced that their mistake was trying to fight Yahweh, a god of the hills, on His home turf. So the king's men told Ben-Hadad to fight the Israelites in the plain where Yahweh, they thought, would have no power. The following spring, Ben-Hadad deployed his army near the city of Aphek, which is thought to be near the present-day kibbutz of Ein Gev on the east shore of the Sea of Galilee.

What is truly amazing is that despite Ahab's devotion to Ba`al—who, we must point out at the risk of stating the obvious, was *the god of the attacking army*—Yahweh again gave victory to Israel to demonstrate His power over the spirit realm.

And the people of Israel were mustered and were provisioned and went against them. The people of Israel encamped before them like two little flocks of goats, but the Syrians filled the country.

And a man of God came near and said to the king of Israel, "Thus says the LORD, 'Because the Syrians have said, "The LORD is a god of the hills but he is not a god of the valleys," therefore I will give all this great multitude into your hand, and you shall know that I am the LORD.'"

And they encamped opposite one another seven days. Then on the seventh day the battle was joined. And the people of Israel struck down of the Syrians 100,000 foot soldiers in one day. And the rest fled into the city of Aphek, and the wall fell upon 27,000 men who were left. (1 Kings 20:27–30a, ESV)

Unfortunately for Ahab, when he learned that Ben-Hadad had been captured in Aphek, he let him go. Now, that sounds enlightened and merciful to us today, from a twenty-first century perspective. But Ahab was in the middle of a holy war and he was fighting for the wrong side.

Ben-hadad also fled and entered an inner chamber in the city. And his servants said to him, "Behold now, we have heard that the kings of the house of Israel are merciful kings. Let us put sackcloth around our waists and ropes on our heads and go out to the king of Israel. Perhaps he will spare your life." So they tied sackcloth around their waists and put ropes on their heads and went to the king of Israel and said, "Your servant Ben-hadad says, 'Please, let me live.'" And he said, "Does he still live? He is my brother."

Now the men were watching for a sign, and they quickly took it up from him and said, "Yes, your brother Ben-hadad." Then he said, "Go and bring him." Then Ben-hadad came out to him, and he caused him to come up into the chariot. And Ben-hadad said to him, "The cities that my father took from your father I will restore, and you may establish bazaars for yourself in Damascus, as my father did in Samaria." And Ahab said, "I will let you go on these terms." So he made a covenant with him and let him go....

So the prophet departed and waited for the king by the way,

disguising himself with a bandage over his eyes. And as the king passed, he cried to the king and said, "Your servant went out into the midst of the battle, and behold, a soldier turned and brought a man to me and said, 'Guard this man; if by any means he is missing, your life shall be for his life, or else you shall pay a talent of silver.' And as your servant was busy here and there, he was gone." The king of Israel said to him, "So shall your judgment be; you yourself have decided it."

Then he hurried to take the bandage away from his eyes, and the king of Israel recognized him as one of the prophets. And he said to him, "Thus says the LORD, '**Because you have let go out of your hand the man whom I had devoted to destruction, therefore your life shall be for his life, and your people for his people.**'" (1 Kings 20:30b–34, 38–42, ESV, emphasis added)

Ben-Hadad was *kherem*, devoted to destruction. But Ahab let him live, thus forfeiting his own life and kingdom. That was the same mistake made by Saul with Agag, king of the Amalekites (see 1 Samuel 15).

That drama plays out a couple chapters later in 1 Kings, in a wonderful scene that depicts the divine council in action.

Three years later, Ahab and King Jehoshaphat of Judah had taken the field to recover Ramoth-Gilead, a Levitical city east of the Jordan River, from the kingdom of Aram. It appears that Ben-Hadad reneged on his deal. Well, Yahweh decreed that Ahab was to die in this battle, but He consulted with the divine council on how to draw Ahab into going to war.

Then the king of Israel gathered the prophets together, about four hundred men, and said to them, "Shall I go to battle against Ramoth-gilead, or shall I refrain?" And they said, "Go up, for the Lord will give it into the hand of the king."

But Jehoshaphat said, "Is there not here another prophet of the LORD of whom we may inquire?" And the king of Israel said to Jehoshaphat, "There is yet one man by whom we may inquire of the LORD, Micaiah the son of Imlah, but I hate him, for he never prophesies good concerning me, but evil." And Jehoshaphat said, "Let not the king say so." Then the king of Israel summoned an officer and said, "Bring quickly Micaiah the son of Imlah."

Now the king of Israel and Jehoshaphat the king of Judah were sitting on their thrones, arrayed in their robes, at the threshing floor at the entrance of the gate of Samaria, and all the prophets were prophesying before them. And Zedekiah the son of Chenaanah made for himself horns of iron and said, "Thus says the LORD, 'With these you shall push the Syrians until they are destroyed.'" And all the prophets prophesied so and said, "Go up to Ramoth-gilead and triumph; the LORD will give it into the hand of the king."

And the messenger who went to summon Micaiah said to him, "Behold, the words of the prophets with one accord are favorable to the king. Let your word be like the word of one of them, and speak favorably." But Micaiah said, "As the LORD lives, what the LORD says to me, that I will speak." And when he had come to the king, the king said to him, "Micaiah, shall we go to Ramoth-gilead to battle, or shall we refrain?" And he answered him, "Go up and triumph; the LORD will give it into the hand of the king." But the king said to him, "How many times shall I make you swear that you speak to me nothing but the truth in the name of the LORD?" And he said, "I saw all Israel scattered on the mountains, as sheep that have no shepherd. And the LORD said, 'These have no master; let each return to his home in peace.'" And the king of Israel said to Jehoshaphat, "Did I not tell you that he would not prophesy good concerning me, but evil?"

And Micaiah said, "Therefore hear the word of the LORD: **I saw the LORD sitting on his throne, and all the host of heaven standing beside him on his right hand and on his left; and the LORD said, 'Who will entice Ahab, that he may go up and fall** at Ramoth-gilead?' And one said one thing, and another said another. Then a spirit came forward and stood before the LORD, saying, 'I will entice him.' And the LORD said to him, 'By what means?' And he said, 'I will go out, and will be a lying spirit in the mouth of all his prophets.' And he said, 'You are to entice him, and you shall succeed; go out and do so.'

Now therefore behold, the LORD has put a lying spirit in the mouth of all these your prophets; the LORD has declared disaster for you." (1 Kings 22:5–23, ESV, emphasis added)

By the way, did you notice how suspicious Jehoshaphat was of the false prophets? There was a consensus of four hundred of them, and Jehoshaphat said, "Isn't there another prophet of Yahweh we can ask?" Jehoshaphat's spiritual discernment told him something wasn't right.

There is no clearer picture of the divine council at work in the Bible than the scene above, except maybe for the first two chapters of Job. Yahweh doesn't lead a pantheon. This is not a polytheistic doctrine. For His own reasons, the beings that He created to serve him play an active role in carrying out His will.

Now, some who reject the divine council view despite the plain text of the Bible argue that God doesn't need a council. That is unquestionably true. But arguing, by extension, that He therefore would not have created one is false. God doesn't need you or me, either, and yet here we are—me writing this book and you reading it.

Micaiah's prophecy was, of course correct. Even though Ahab ordered Micaiah arrested and held in Samaria until his return, the king knew something was up and tried to come up with a way to short-circuit the prophecy.

So the king of Israel and Jehoshaphat the king of Judah went up to Ramoth-gilead. And the king of Israel said to Jehoshaphat, "I will disguise myself and go into battle, but you wear your robes." And the king of Israel disguised himself and went into battle. (1 Kings 22:29–30, ESV)

Nice. "Wear your royal robes while I dress like a commoner so they shoot at you." And Jehoshaphat was, in fact, nearly killed. The Syrians mistook him for Ahab for a while, but finally turned back when they realized their mistake. But an archer, shooting at random, put an arrow through a chink in Ahab's armor, and he never made it back to Samaria alive.

Let this be a lesson: Don't call a guy who worships the Enemy your brother.

Ahab had no excuse. He should have known better. Yahweh had given him two astonishing victories over the Arameans, victories that were clearly supernatural. Yahweh even led prophets to tell Ahab the outcomes of the battles beforehand. And both miraculous victories happened *after* the supernatural showdown that's the at the heart of this chapter.

Let's go back to 1 Kings 18. This is the famous challenge on Mount Carmel between the prophet Elijah and the priests of Ba`al.

It hadn't rained in Israel for three years and a severe famine gripped the land. Ahab, no doubt at the urging of Jezebel, had hunted all over the Near East for Elijah. We can assume the prophet was hidden from the prying eyes of Ahab's spies by Yahweh.

Things were grim for those who were faithful to God. Jezebel had "cut off" the prophets of Yahweh, a euphemism that means terminated with extreme prejudice. Ahab's chief of staff, Obadiah, had secretly hidden a hundred of the remaining prophets in two caves and kept them alive by sneaking bread and water to them.

Then, without warning, Elijah presented himself to Obadiah—who was understandably nervous. It might have appeared, to Jezebel if not to Ahab, that Obadiah might have conveniently overlooked the prophet's hiding place. But Elijah was good to his word and stayed put until Obadiah returned with Ahab.

You have probably heard the story: Elijah did spiritual battle with 450 prophets of Ba`al and smoked them—literally—when Yahweh's heavenly fire consumed his sacrifice.

Even this story, which is as obvious a showdown between two elohim as there is in the Bible, has been drained of much of the supernatural element by our churches today. Most of us have been taught that the prophets of Ba`al got no answer because there was nothing to answer them, the poor, deluded fools.

That's where we need to correct the record.

Those supernatural entities do exist. Please review Deuteronomy 32:8–9 (using the ESV, ISV, or the NET Bible), the verses that explain how Yahweh divided the nations after the Tower of Babel, allotting them to those seventy small-G gods. Then re-read Exodus 12:12, where God told Moses that on the night of the Passover He would execute judgments "on all the gods of Egypt." That wasn't a figure of speech.

Look, if this worldview is incorrect, then the apostle Paul was either lying or badly misinformed when he repeatedly warned us against principalities, powers, thrones, dominions, rulers, and the elemental spirits of the world.

To put it plainly, the three-year prelude to the showdown at Mount Carmel was all about humiliating Ba`al in front of the nation of Israel. Ba`al, you remember, was not just the storm-god; he was the god of life-giving rain essential for crops, herds, and humans. And for three years, Ba`al hadn't delivered.

To be precise, Jesus' half-brother James wrote that the rain had been stopped for three and a half years. Forty-two months.

Gee. Where else do we see that specific period of time in the Bible?

Then I was given a measuring rod like a staff, and I was told, "Rise and measure the temple of God and the altar and those who worship there, but do not measure the court outside the temple; leave that out, for it is given over to the nations, and they will trample the holy city for forty-two months. **And I will grant authority to my two witnesses, and they will prophesy for 1,260 days, clothed in sackcloth.**"

These are the two olive trees and the two lampstands that stand before the Lord of the earth. And if anyone would harm them, fire pours from their mouth and consumes their foes. If anyone would harm them, this is how he is doomed to be killed. **They have the power to shut the sky, that no rain may fall during the days of their prophesying,** and they have power over the waters to turn them into blood and to strike the earth with every kind of plague, as often as they desire. (Revelation 11:1—6, ESV, emphasis added)

So Mount Carmel was to humiliate Ba`al *and* foreshadow end-times prophecy. Now, by this point in the book, do you think for a moment that was just a coincidence? This is one of the reasons some prophecy scholars believe Elijah will be one of the two witnesses of Revelation.

To set the stage: The confrontation on Mount Carmel was a national event. Ahab invited all of Israel, the northern kingdom, and he gathered not just the 450 prophets of Ba`al but the 400 prophets of Asherah who ate at Jezebel's table as well.

Considering that they didn't have cable, satellite, or Internet in Israel 2,875 years ago, who wouldn't have taken a day off to see this? Especially after the god brought to the land by the foreign-born queen hadn't watered their crops in three years? Hey, the people in my home state of Illinois came from miles around in 1858 to see Abe Lincoln and Stephen Douglas do nothing but *talk* for two hours. Can you imagine the buildup for an actual battle between gods? This was a supernatural Super Bowl! And that's exactly the point of this event at this particular location.

Mount Carmel was considered holy for at least six hundred years before Elijah's day. The name Carmel means "vineyard of God"—or, considering the influence of Amorite/Canaanite religion (especially under Ahab and Jezebel), "vineyard of El." Pharaoh Thutmose III, on his way to the Battle of Megiddo mentioned earlier, probably meant Mount Carmel in an inscription that mentioned *Rash-Qadesh*, or "holy headland."

Later, in the fourth century B.C., a Greek geographer called Mount Carmel "the mountain of Zeus," specifically an incarnation of Zeus called Zeus Heliopolitanus. That was a reference to Heliopolis in Phoenicia, located at the north end of Lebanon's Bekaa valley. We know that city today by a different name: Baalbek.

If you've researched megalithic structures at all, you've heard the name Baalbek. We won't go into a discussion of the Trilithon, the three massive stones from the retaining wall of the temple of Jupiter. Let's just say that at 880 tons each, they're impressive.[33] The important point is that Jupiter's temple at Baalbek was built on top of an older temple to Hadad—Ba`al. Remember, Jupiter = Zeus = Ba`al.

So Mount Carmel was a site holy to Ba`al. In the time of Ahab, it marked the southern boundary of Phoenicia, a kingdom devoted to the worship of Ba`al. After all, Phoenicia's northern border in Elijah's day was Mount Zaphon, the site of Ba`al's palace.

Put another way, since the days of David and Solomon, Ba`al had been moving the boundary stone to encroach on Yahweh's territory. This could not stand.

Finally, the big day arrived at Carmel. The Bible records that there was quite a crowd. It must have been a great show.

> So Ahab sent to all the people of Israel and gathered the prophets together at Mount Carmel. And Elijah came near to all the people and said, "How long will you go limping between two different opinions? If the LORD is God, follow him; but if Baal, then follow him." And the people did not answer him a word. Then Elijah said to the people, "I, even I only, am left a prophet of the LORD, but Baal's prophets are 450 men. Let two bulls be given to us, and let them choose one bull for themselves and cut it in pieces and lay it on the wood, but put no fire to it. And I will prepare the other bull and lay it on the wood and put no fire to it. And you call upon the name of your god, and I will call upon the name of the LORD, and the God who answers by fire, he is God." And all the people answered, "It is well spoken."
>
> Then Elijah said to the prophets of Baal, "Choose for your-selves one bull and prepare it first, for you are many, and call upon

the name of your god, but put no fire to it." And they took the bull that was given them, and they prepared it and called upon the name of Baal from morning until noon, saying, "O Baal, answer us!" But there was no voice, and no one answered. And they limped around the altar that they had made.

And at noon Elijah mocked them, saying, "Cry aloud, for he is a god. Either he is musing, or he is relieving himself, or he is on a journey, or **perhaps he is asleep and must be awakened."** And they cried aloud and **cut themselves after their custom with swords and lances, until the blood gushed out upon them.** And as midday passed, they raved on until the time of the offering of the oblation, but there was no voice. No one answered; no one paid attention. (1 Kings 18:20–29, ESV, emphasis added)

The prophets of Ba`al must have thought they had this in the bag. After all, Ba`al was the storm-god. His weapon was the thunderbolt. Easy, right?

Eh—not so much.

The behavior of Ba`al's priests was apparently typical for his cult. An Egyptian papyrus written about two hundred years before the time of Elijah, *The Story of Wen-Amun*, includes a description of a frenzied prophet of Ba`al. We also know that by Elijah's day, Ba`al had joined the growing ranks of "dying and rising gods" in the ancient Near East. Deities such as Tammuz, Adonis, Attis, and Dionysus were mourned and celebrated every year as part of the cycle of agriculture as the land turned brown and then green again. That explains Elijah's comment, "Perhaps he is asleep and must be awakened."

A tablet from Ugarit suggests that the Bible didn't exaggerate what must have been a bloody scene on Mount Carmel. In the early days of the world, it was believed that Ba`al not only had to defeat Yam the sea-god, the "darling of El." he had a pair of battles with the "beloved of El." Mot, the god of death. After Mot vanquished Ba`al in round one, the gods mourned. And, setting an example that's been followed by cults through the ages, a lot of blood was spilled.

[El] poured straw of mourning on his head,
dust of wallowing on his crown;
for clothing he covered himself with sackcloth;
he scraped (his) skin with a stone,
with a flint for a razor
he shaved (his) side-whiskers and beard;
he harrowed his collar-bone,
he ploughed (his) chest like a garden,
he harrowed (his waist) like a valley.
He lifted up his voice and cried:
"Baal is dead!"
(CTA 5 v 11–23; CML, 73, emphasis added)

It's hard to imagine from the biblical text what it must have looked like that day on Mount Carmel, but you don't have to look very hard on the web to get a good idea. Search sometime for pictures of the Shia Muslim festival called Ashura, which marks the death of Husayn ibn Ali, the grandson of Muhammad, at the Battle of Karbala in A.D. 680. But be warned: The photos you'll find will be gruesome.

Those images, and the scene on Mount Carmel that fateful day circa 860 B.C., illustrate in living color the difference between the Enemy and our Lord, Jesus Christ: They want us to shed our blood for them. We're saved from eternal death because Christ shed *His* blood for *us*.

Well, after the priests of Ba`al struck out in their turn at bat, Elijah stepped up to the plate.

Then Elijah said to all the people, "Come near to me." And all the people came near to him. And **he repaired the altar of the LORD that had been thrown down.** Elijah took twelve stones, according to the number of the tribes of the sons of Jacob, to whom the word of the LORD came, saying, "Israel shall be your name," and with the stones he built an altar in the name of the LORD. And he made a trench about the altar, as great as would contain two seahs of seed. And he put the wood in order and cut the bull in pieces and laid it on the wood. **And he said, "Fill four jars with water and pour it on the burnt offering and on the wood." And he said, "Do it a second time." And they did it a second time. And he said, "Do it a third time." And they did it a third time. And the water ran around the altar and filled the trench also with water.**

And at the time of the offering of the oblation, Elijah the prophet came near and said, "O Lord, God of Abraham, Isaac, and Israel, let it be known this day that you are God in Israel, and that I am your servant, and that I have done all these things at your word. Answer me, O Lord, answer me, that this people may know that you, O Lord, are God, and that you have turned their hearts back." Then **the fire of the Lord fell and consumed the burnt offering and the wood and the stones and the dust, and licked up the water that was in the trench.** And when all the people saw it, they fell on their faces and said, "The Lord, he is God; the Lord, he is God." (1 Kings 18:30–39, esv, emphasis added)

Because you're alert, you've noticed that there had previously been an altar to Yahweh on Mount Carmel that had been thrown down. As we said, Ba`al was encroaching on Yahweh's turf.

Now, did you catch that Elijah dug a trench around the altar? Converting from seahs to liquid measure, it was about big enough to hold eight gallons of water. Then he had enough water poured on the altar to fill the trench. Remember, this was after *three years without rain*. Elijah wasn't wasting scarce, precious water just for theatrical effect. The prophet was underlining a point: Ba`al had no power in the land except what was allowed by Yahweh.

It was also symbolically significant. Four jars of water poured three times each equaled twelve jars, one for each of the tribes of Israel. You can imagine the murmuring in the crowd at all that water being poured out. Remember, the land was so dry even king Ahab was struggling to feed his animals.

You know what happened next. But it went down maybe a little differently than the way we've been told. While most of the artists' renderings portray the "fire of the LORD" coming from the sky like a blast from a heavenly flamethrower, scholars generally agree that what's described in 1 Kings 18:38 is lightning. That makes sense. After all, the three-year prelude to this challenge was a demonstration that Yahweh controlled the life-giving rain, not Ba`al. The icing on this epic battle was that Yahweh consumed the sacrifice by wielding the signature weapon of Ba`al—the thunderbolt—when Ba`al could not.

To put this in our Hollywood-centered, media-saturated, twenty-first-century context: This was like a pro wrestling heavyweight championship match where the cocky young challenger got beat with his own patented finishing move.

Not to put too fine a point on it, but this confrontation wasn't entirely about Elijah trusting God. It wasn't even *mainly* about Elijah trusting God. The encounter on Mount Carmel was included in the Bible because it was a powerful demonstration to the people—and especially to the rebellious Fallen—that Israel was the inheritance of Yahweh and Yahweh alone. And to make sure everybody understood, Yahweh orchestrated the showdown to emphasize His power over what was supposed to be the domain of the lesser gods—just as He did at in the Valley of Aijalon, at Jericho, and at the Red Sea.

The Israelites got the message. "[T]hey fell on their faces and said, 'The LORD, He is God; the LORD, He is God.'"

Things ended badly for the prophets of Ba`al. At Elijah's orders, the people killed all 450 of them. No word in the text about what happened to the prophets of Asherah. Maybe they ran away. The text doesn't make it clear that they even showed up. But the outcome was devastating to Jezebel. The next chapter of 1 Kings opens with the queen swearing to the gods that Elijah was a dead man.

Ahab didn't seem too bothered; he ate a meal on Mount Carmel before heading back to Jezreel in his chariot. (Maybe he wasn't in a hurry to bring Jezebel the news.) To put a cap on the day, the king hadn't even finished his meal before Yahweh sent a downpour so heavy that Elijah warned the king to get going "lest the rain stop you."

And just like the soldiers of Sisera four centuries earlier, the 450 priests of Ba`al were swept away by the waters of Kishon.

———

One more thing. In news broadcasting, we call this the kicker—the story you save for last because of the "wow" factor.

It's probable that the Ba`al worshiped by Jezebel and the 450 prophets who died on Carmel was not the Semitic storm-god, Hadad. At that point in history, the chief god of the city-state of Tyre was Melqart. His name, based on the roots *mlk qrt*, meant "king of the city."

Melqart was introduced to Tyre back in the days of David and Solomon. Hiram, who supplied material and workmen for the Temple in Jerusalem, and his father, Abibaal ("my father is Ba`al"), apparently wanted to break the power of the priests of Ba`al by making their lord play second fiddle to a new god whose priests answered to the king. So they introduced Melqart, whose cult grew with the influence of Tyre, especially as the Phoenician city-state planted successful colonies around the Mediterranean. The god was worshipped as far away as Cadiz in Spain and as late as the first century A.D.

Tyre's most famous colony, Carthage, worshiped Melqart as its chief god until the early fifth century B.C., when Carthage asserted its independence from Tyre and elevated Ba`al Hammon to the top spot. That's the god that horrified the Romans for accepting child sacrifice. (Despite what you may have heard from well-meaning pastors or teachers, there is no evidence that Ba`al-Hadad or any other Near Eastern storm-god like Zeus, Jupiter, Addu, Teššub, Tarhunt, Teisheba, or Iškur was worshiped through human sacrifice, Jeremiah 19:5 notwithstanding. It's likely Jeremiah was referring to Moloch and/or Chemosh, the gods of Ammon and Moab.)[34]

Short rabbit trail: We should mention that Ba`al Hammon was identified by the Greeks and Romans as Kronos/Saturn, the god who ate his children during the so-called golden age before the flood. If our theory is correct that the Phoenicians were the physical and spiritual descendants of the Amorites, it shouldn't surprise us that the Carthaginians made the king of the Titans (the Didanu/Tidanu) their god. And that's why the occult working Tom Horn described in *Zenith 2016* to bring back the golden age of "old Saturn's reign" should concern you.

The point is that it's a pretty good bet the priests facing Elijah on Mount Carmel that day served Melqart rather than Ba`al-Hadad. His cult may have been introduced to Israel by Jezebel, but he may have arrived even earlier. The second-century Christian theologian Tatian quotes Phoenician historians who wrote that Hiram gave one of his daughters to Solomon as a wife, and through her Solomon began to worship Astarte, the goddess served by Jezebel's father. So the seed of the Melqart cult could have been planted in Israel more than a hundred years before the show-down on Mount Carmel.

As we mentioned, Melqart was one of the dying and rising gods. He would die in the fall and rise again every spring, following the annual growth cycle of agriculture. Religious rites accompanied the god's death and resurrection every year. The Greek historian Herodotus visited Tyre in the fifth century B.C. and reported that he'd seen Melqart's tomb inside his magnificent temple. That would be consistent with Elijah's jibe that the god was "asleep and must be awakened."

On the other hand, and here's the kicker, the presence of a tomb *might* indicate that the cult of Melqart was based on a deified person instead of a spirit. You see, Melqart was the Phoenician name of Heracles—better known as Hercules.

Digest that for a moment. In all probability, the 450 priests who met their doom on Mount Carmel that fateful day served *Hercules.*

Assuming he was based on an historic character, could Melqart/ Hercules have been an actual demigod? Possibly. The Watchers and the Nephilim were real, so Hercules may well have been one of the "mighty men who were of old."

And since the Phoenicians of Tyre descended from the Amorites, who venerated and summoned spirits they thought were their glorious ancestor kings (the council of the Didanu/Titans), a Phoenician religion based on one those demonic spirits is actually, from their perspective, logical.

But we can be sure of this: If he lived, Hercules was nothing like the cartoon character who's been entertaining kids for the last half a century.

———

Before we leave this section, let's fit the Mount Carmel incident into the timeline of the long war between Yahweh and the Fallen. In other words, let's examine the incident and references to the players elsewhere in the Bible to see what relevance it might have for where we're going.

You'll remember from the first part of this book that the prophet Ezekiel had some unpleasant words from Yahweh for the king of Tyre. Considering what we've just discussed regarding the identity of the god Yahweh shamed on Mount Carmel, some of the writings of the prophet Ezekiel take on a new dimension.

We earlier examined Ezekiel 28, which, while it's described as a lament over the king of Tyre, most scholars agree is directed at the nachash, the serpentine supernatural being who rebelled against the authority of Yahweh in Eden. As a result, he was thrown down from the mountain of God to serve as lord of the dead. Quite a fall from being "perfect in beauty."

But if we back up a chapter to Ezekiel 27, the passage takes on a different flavor if the prophet was addressing the supernatural king of the city, Melqart.

Tyre founded the greatest commercial empire of the ancient world. Maybe they had to—the thin strip of fertile land between the coast and the Lebanon mountains doesn't allow for much expansion inland. By building ships and sailing the Mediterranean and beyond, Tyre became an international power through trade, as Venice did in the Middle Ages.

The Phoenicians of Tyre, descendants of the Amorites/Canaanites of the second millennium B.C., established powerful colonies in Spain and North Africa that ventured into the Atlantic and reached the coasts of West Africa and Britain. Carthage, led by the great general Hannibal, nearly brought Rome to its knees in the third century B.C.

Ezekiel 27 reveals interesting details about the city that worshiped Hercules, and possibly suggests a new aspect of analyzing the great apostate religion that will emerge in the end times.

> Your borders are in the heart of the seas;
> your builders made perfect your beauty.
> **They made all your planks**
> **of fir trees from Senir;**
> **they took a cedar from Lebanon**
> **to make a mast for you.**
> **Of oaks of Bashan**
> **they made your oars;**
> they made your deck of pines [...]
> Of fine embroidered linen from Egypt
> was your sail,
> serving as your banner;
> **blue and purple from the coasts of Elishah**
> **was your awning.**
> The inhabitants of Sidon and Arvad
> were your rowers;
> your skilled men, O Tyre, were in you;
> they were your pilots.
> The elders of Gebal and her skilled men were in you,
> caulking your seams;
> all the ships of the sea with their mariners were in you
> to barter for your wares.
> (Ezekiel 27:4–8, ESV, emphasis added)

We have here an intriguing spiritual analysis of the history of Tyre. Ezekiel likened Tyre to a ship on the ocean—the "heart of the seas"—and then specified the sources of the materials used to build the ship. It's an interesting metaphor on several levels.

First, "the heart of the seas" is a phrase Ezekiel repeats in chapter 28:

Son of man, say to the prince of Tyre, "Thus says the Lord GOD:
'Because your heart is proud,
and you have said, "I am a god,
I sit in the seat of the gods,
in the heart of the seas,"
yet you are but a man, and no god,
though you make your heart like the heart of a god.'"
(Ezekiel 28:2, ESV, emphasis added)

That verse refers to Yahweh's holy mountain, Eden. It's similar to a phrase from Amorite/Canaanite myth used to describe another "seat of the gods." the mountain of El's abode, which was found "at the sources of the two rivers in the midst of the fountains of the double-deep."[35]

This also evokes the image of Enki's ancient temple at Eridu, the *E-abzu*, the House of the Abyss. Some scholars identify El as Enki on that basis, although the character of El in Canaanite myth is not much like Enki/Ea in the Sumerian, Akkadian, and Babylonian myths. But file a mental bookmark at this connection anyway, because we'll revisit the idea in the next chapter.

To make sure we catch the connection, Ezekiel writes that the planks of the ship-of-Tyre come from Senir. That's the Amorite name for Mount Hermon, El's mount of assembly. The oars, which propel a ship, were made from oaks of Bashan. You'll remember that Bashan, at the foot of Mount Hermon, was the kingdom of Og, last of the Rephaim, whose land was known as the gateway to the netherworld.

In other words, viewing this through the divine council paradigm, Ezekiel expressly connects Tyre to the Watchers/Titans, and by extension their children, the Nephilim/Rephaim, by describing Tyre's very foundation and *raison d'être* as coming straight from Mount Hermon, the site of the Watchers' rebellion in Genesis 6. Boiling it down, it's all about "the iniquity of the Amorites," the people who founded the occult system of Babylon. It's not a reference alert readers in the sixth century B.C. would have missed.

And remember, Ezekiel wrote his prophecies while living as an exile in the heart of Babylonia. This stuff was very top-of-mind for the prophet.

Ezekiel's references to Egypt and the coasts of Elishah, probably Cyprus (known in ancient times as Alashiya), were possibly a reference to Tyre's status as a merchant kingdom, or it might be an allusion to lands where Ba`al and Astarte were (or had been) among the chief deities in the pantheon. Astarte (Ishtar/Inanna) was brought to Egypt with Ba`al and Resheph by the Hyksos about nine hundred years before Elijah's encounter on Mount Carmel, and they were transplanted to Cyprus when Tyre established a colony there in Elijah's day.

The "blue and purple from the coasts of Elishah" was the dye for which Tyre became famous in the classical world. It was produced throughout the eastern Mediterranean from the murex, a type of sea snail.

The "crew" of the city—Sidon, Arvad, and Tyre—were, like the Amorites before them, names of descendants of the cursed Canaan. That was no coincidence. It's also unlikely that the parallel between Ezekiel's prophecy over Tyre and an end times prophecy of Babylon is coincidental.

Your rowers have brought you out
into the high seas.
**The east wind has wrecked you
in the heart of the seas.**
Your riches, your wares, your merchandise,
your mariners and your pilots,
your caulkers, your dealers in merchandise,
and all your men of war who are in you,
with all your crew
that is in your midst,
sink into the heart of the seas

on the day of your fall.
At the sound of the cry of your pilots
the countryside shakes,
and down from their ships
come all who handle the oar.
The mariners and all the pilots of the sea
stand on the land
and shout aloud over you
and cry out bitterly.
They cast dust on their heads
and wallow in ashes;
they make themselves bald for you
and put sackcloth on their waist,
and they weep over you in bitterness of soul,
with bitter mourning.
In their wailing they raise a lamentation for you
and lament over you:
"Who is like Tyre,
like one destroyed in the midst of the sea?"
(Ezekiel 27:26–32, ESV, emphasis added)

Compare that with this prophecy from the Revelation of John:

After this I saw another angel coming down from heaven, having
great authority, and the earth was made bright with his glory.
And he called out with a mighty voice,
"Fallen, fallen is Babylon the great!
She has become a dwelling place for demons,
a haunt for every unclean spirit,
a haunt for every unclean bird,
a haunt for every unclean and detestable beast....
Alas, alas, for the great city
that was clothed in fine linen,

in purple and scarlet,
adorned with gold,
with jewels, and with pearls!
For in a single hour all this wealth has been laid waste."
And all shipmasters and seafaring men, sailors and all whose
trade is on the sea, stood far off
and cried out as they saw the smoke of her burning,
"What city was like the great city?"
And they threw dust on their heads as they wept and mourned,
crying out,
"Alas, alas, for the great city
where all who had ships at sea
grew rich by her wealth!"
For in a single hour she has been laid waste.
(Revelation 18:1–2, 15–19, ESV, emphasis added)

Please note: Babylon was located about 260 miles from the head of the Persian Gulf. While it was certainly wealthy in its heyday, which ended about four hundred years before John the Revelator, it was never a port city.

It could be the references to seafaring men and the ships at sea are at least partly metaphorical, allusions to the "heart of the seas" on El's rebellious mount of assembly. Is it there, in a spiritual conflict mostly hidden from human sight, that the end-times Babylon will founder? It's significant that it was the east wind that wrecked Tyre in the heart of the seas; the east wind is what blasted the grain in Joseph's dream of a coming famine, and it's what Yahweh used to bring the plague of locusts to Egypt and to part the Red Sea.

The east wind was also Yahweh's instrument of destruction to destroy the ships of another seagoing nation:

Great is the LORD and greatly to be praised
in the city of our God!

His holy mountain,
beautiful in elevation,
is the joy of all the earth,
Mount Zion, in the far north,
the city of the great King.
Within her citadels God
has made himself known as a fortress.
For behold, the kings assembled;
they came on together.
As soon as they saw it, they were astounded;
they were in panic; they took to flight.
Trembling took hold of them there,
anguish as of a woman in labor.
By the east wind you shattered
the ships of Tarshish.
As we have heard, so have we seen
in the city of the LORD of hosts,
in the city of our God,
which God will establish forever. *Selah.*
(Psalm 48:1–8, ESV, emphasis added)

In other words, the east wind, which comes from the direction of the entrances to Eden, the tabernacle, and the Temple, is a wind of divine judgment.

The passages in Ezekiel 27 and Revelation 18 refer to the same future event. While Tyre was destroyed by Alexander the Great, and Babylon was left deserted by the infighting between his successors, those devastations were already-but-not-yet fulfillments.

Remember, the Amorite kings of old Babylon, Assyria, and Ugarit, and possibly their Phoenician descendants, too, believed their ancestors were the "mighty men who were of old," the Rephaim/Nephilim, children of the Didanu/Titans/Watchers. The Amorites earned special condemnation from Yahweh for bringing back to the post-Flood world forbidden secret knowledge. The religion of Babylon glorified occult practices such as sorcery, necromancy, and sex magic, encouraged and embodied by Inanna (and her later incarnations Ishtar, Astarte, Aphrodite, and Venus) and the priesthood of Marduk during the annual *akitu* ritual.

While it's beyond the scope of this book to document those occult practices down through the ages, it appears that through their contact with the Greek and Roman world, the descendants of the Amorites, the Phoenicians, preserved and spread to the secret societies of Europe and the West the occult spirit of Babylon and a desire to bring back the Golden Age of Kronos—the age of the Watchers and the Nephilim.

SUMMING UP

As with the crossing of the Red Sea and the confrontation at Jericho, the showdown on Mount Carmel between Elijah and the prophets of Ba`al has supernatural implications that are left out of most accounts we hear at church. Ba`al, or the entity that allowed itself to be called "lord" by its Amorite/Canaanite devotees, is a real god. His attempt to encroach on Yahweh's land, Israel, was the reason for the public contest on a mountain that had formerly held an altar to Yahweh—one that had been torn down to put up a new one for Ba`al.

The setting and the style of the confrontation, following three and a half years without rain in Israel, was specifically to demonstrate that Ba`al had no power in the land. Yahweh not only prevented Ba`al from performing his main function, bringing life-giving rain, He consumed the sacrifice prepared by Elijah with Ba`al's signature weapon, a thunderbolt.

Even after seeing this with his own eyes, and two miraculous victories over much larger Aramean armies, king Ahab did not repent of his worship of Ba`al. He chose his wife, Jezebel, over the Lord of Hosts, and Ahab paid the ultimate price. His rebellion cost Ahab his life and his kingdom. And despite the reprieve, the continued apostasy of the northern kingdom finally provoked Yahweh to withdraw His protection from Israel.

Samaria was captured by the Assyrians in 722 B.C., and the northern kingdom ceased to exist as an independent power.

We also examined the undeniable parallels between the lament over Tyre in Ezekiel 27 and the lament of the kings of the earth over the destruction of Babylon in Revelation 18. We will consider them further and try to identify the mysterious Babylon of end-times prophecy.

7

ZION

ONCE AGAIN, WE'RE going to fast-forward through history. We'll skip over some really rich material from a warfare-in-the-heavenlies perspective, but the original plan for this book was to focus on holy mountains as key conflict zones in the long war between the Fallen and Yahweh. Hopefully, this isn't the last book inside your humble author's head.

The rest of the Old Testament is chock full of supernatural war that's been bleached out of what we hear in church. We've already mentioned some of it—Daniel's writing on the wall, for example—and some will follow in this section. But there's a gold mine in some of the harder-to-understand sections of Job, Ezekiel, Daniel, Isaiah, and the minor prophets that we'll come back to someday, God willing.

As we said at the outset, the ultimate goal in the long war is to establish the dominant mount of assembly in the cosmos. That was the mission of the divine rebel in Eden.

You said in your heart,
"I will ascend to heaven;
above the [*angels*] of God

I will set my throne on high;
I will sit on the mount of assembly
in the far reaches of [*Zaphon*];
I will ascend above the heights of the clouds;
I will make myself like the Most High."
(Isaiah 14:13–14, ESV, bold and italics added)

While the mission of the Fallen is to establish and control the pre-eminent mount of assembly, the prize in the Long War is the mountain Yahweh claimed as His own, the Temple Mount in Jerusalem.

As mountains go, Zion is not all that impressive. At 2,430 feet, it's not even the highest point in Jerusalem. In fact, the mountain called Zion today isn't even the original Mount Zion. The modern Zion is outside the walls of the old city, a peak called the Western Hill in ancient days. It's about eighty feet higher than the Temple Mount, so apparently first-century Jews thought it was a more fitting site for the long-lost palace of King David.

From the time of Abraham onward, the Fallen had a pretty good idea that Mount Moriah, later Mount Zion, was Yahweh's future home on earth. He telegraphed the move when He called Abraham there to test his faith. That was such a significant event in the life of Abraham that the Fallen couldn't help but notice. For starters, it was three days' travel from their camp to the mountain, probably a twenty or thirty-mile journey, far enough (and possibly dangerous enough) that Abraham took retainers with them—the "young men" of Genesis 22:5.

> And Isaac said to his father Abraham, "My father!" And he said, "Here am I, my son." He said, "Behold, the fire and the wood, but where is the lamb for a burnt offering?" Abraham said, "God will provide for himself the lamb for a burnt offering, my son." So they went both of them together. (Genesis 22:8, ESV)

A dynamic young preacher this author once heard made an astute observation about the difference between Abraham and today's church.

In Abraham's case, he had the fire and he had the wood, so Isaac asked, "Where is the lamb?" In contrast, today's church in America has the Lamb, and we have the wood (the cross), but *where is the fire?* But I digress.

If you know your Old Testament, you know that the mountain did not come into Israel's possession when Joshua led the conquest. Although the Israelites defeated the coalition led by Jerusalem's king Adoni-zedek during Joshua's Long Day, the city itself remained in Amorite hands. It fell to David, almost four hundred years later, to take the city (1 Chronicles 11). From the time of David and Solomon, it would remain under the control of the kingdom of Judah until Nebuchadnezzar made an end of the independent Jewish state in 586 B.C.

The Hebrew prophets knew that Zion was where Yahweh made His home on earth. They also knew about the competing claims of Zaphon and Hermon. If you read the Old Testament carefully, you'll find several obvious polemics written by the prophets against those mountains and the gods who called them home.

Some of the references are stunning. For example, did you know that Yahweh led a military campaign against Mount Hermon?

> O mountain of God, mountain of Bashan;
> O many-peaked mountain, mountain of Bashan!
> Why do you look with hatred, O many-peaked mountain,
> at the mount that God desired for his abode,
> yes, where the LORD will dwell forever?
> (Psalm 68:15–16, ESV)

The many-peaked mountain of Bashan is Mount Hermon. Hermon is a mountain cluster at the southern end of the Anti-Lebanon range, hence "many-peaked."

Verse 15 looks odd because Mount Hermon is not, and never was, the mountain of God. But the Hebrew words *har elohim* can also be translated "mountain of *the gods,*" which makes more sense in this context. Read it again with the alternate translation of verse 15:

O **mountain of the gods**, mountain of Bashan;
O many-peaked mountain, mountain of Bashan!
Why do you look with hatred, O many-peaked mountain,
at the mount that God desired for his abode,
yes, where the LORD will dwell forever?
(Psalm 68:15–16, ESV, alternate translation)

It changes the entire sense of the psalm—in fact, with the change, the psalm now makes sense.

Verse 16 asks why Hermon looks at the true mountain of God with jealousy and hatred. That's an easy question to answer—Hermon, once the abode of the gods, is now nothing more than a historic landmark—a place where rebellious sons of God planned insurrection, but has since fallen to the army of Yahweh. Mount Hermon is conquered territory in the supernatural war of the ages.

The chariots of God are twice ten thousand,
thousands upon thousands;
the LORD is among them; Sinai is now in the sanctuary.
You ascended on high,
leading a host of captives in your train
and receiving gifts among men,
even among the rebellious, that the LORD God may dwell there.
Blessed be the Lord,
who daily bears us up;
God is our salvation. Selah
Our God is a God of salvation,
and to GOD, the Lord, belong deliverances from death.
But God will strike the heads of his enemies,
the hairy crown of him who walks in his guilty ways.
The Lord said,
"I will bring them back from Bashan,

I will bring them back from the depths of the sea,
that you may strike your feet in their blood,
that the tongues of your dogs may have their portion from the foe."
(Psalm 68:17–23, ESV, emphasis added)

This is a glimpse of war in the heavenlies. These verses describe an assault on Mount Hermon by Yahweh and his heavenly army, possibly right after the verses in Habakkuk 3 when Yahweh went out from Mount Paran (Sinai) and "crushed the head of the house of the wicked."

The parallel of Bashan and the depths of the sea in verse 22 may be references to Yahweh's supernatural enemies. Bashan, gateway to the netherworld and the "place of the serpent." represents the divine rebel from Eden, the nachash. The depths of the sea represent the forces of chaos, made manifest in Leviathan, that Yahweh defeated in an earlier age. It also reinforces our analysis in the previous chapter, linking Ezekiel 27 and the lament over Tyre's destruction "in the heart of the seas" to Hermon, El's mount of assembly.

But the Fallen didn't give up Mount Zion without a fight. Psalm 78 shows that Yahweh took the mountain by force.

He struck down every firstborn in Egypt,
the firstfruits of their strength in the tents of Ham.
Then he led out his people like sheep
and guided them in the wilderness like a flock.
He led them in safety, so that they were not afraid,
but the sea overwhelmed their enemies.
And he brought them to his holy land,
to the mountain which his right hand had won.
He drove out nations before them;
he apportioned them for a possession
and settled the tribes of Israel in their tents.
(Psalm 78:51–55, ESV, emphasis added)

Now, let's look again at the beginning of Psalm 48, which we cited in the last chapter. Because you're paying attention, you noticed that the psalm includes a clear reference to Mount Zaphon:

> Great is the LORD and greatly to be praised
> in the city of our God!
> His holy mountain, beautiful in elevation,
> is the joy of all the earth,
> **Mount Zion, in the far north,**
> **the city of the great King.**
> (Psalm 48:1–2, ESV, emphasis added)

The psalmist is equating Zion with Zaphon as a literary device. Since Zion isn't much of a mountain and it sure isn't in the far north, a deeper meaning is implied. This is a comparison that takes the imagery of Ba`al's mount of assembly, his *har mo'ed*, and applies it to Zion to show the superiority of Yahweh and His mount of assembly.

It's clear from the time of David onward that the psalmists and prophets understood the significance of Zion. It was Yahweh's home on earth, where He would rule and reign forever.

> The LORD reigns; let the peoples tremble!
> He sits enthroned upon the cherubim; let the earth quake!
> The LORD is great in Zion;
> he is exalted over all the peoples....
> Exalt the LORD our God,
> and worship at his holy mountain;
> for the LORD our God is holy!
> (Psalm 99:1–2, 9, ESV)

After all the miracles witnessed by the people of Israel, you'd think they wouldn't need reminders about what happens to those who reject Yahweh. But then, we humans are a stubborn bunch.

When you cry out, let your collection of idols deliver you!
The wind will carry them off,
a breath will take them away.
But he who takes refuge in me shall possess the land
and shall inherit my holy mountain. (Isaiah 57:13, ESV)

By the time of Isaiah, the people in the kingdom of Judah had turned to worshiping many of the rebellious gods. It wasn't just Ba`al and Asherah who drew the Israelites away from Yahweh:

But you who forsake the LORD,
who forget my holy mountain,
who set a table for Fortune
and fill cups of mixed wine for Destiny,
I will destine you to the sword,
and all of you shall bow down to the slaughter,
because, when I called, you did not answer;
when I spoke, you did not listen,
but you did what was evil in my eyes
and chose what I did not delight in.
(Isaiah 65:11–12, ESV, emphasis added)

What makes this passage interesting is not just the reference to God's holy mountain, but the fact that the ESV translators recognized something that led them to capitalize the words Fortune and Destiny. You see, Gad (Fortune) and Meni (Destiny) were gods, and Jews of Isaiah's day were serving up cultic banquets for them.

Gad was the god of good luck. He was worshiped by the Amorites and their descendants, the Canaanites and Phoenicians, from pre-Israelite days down to the fourth or thirdcentury B.C. In spite of the gender difference, Gad was equivalent to the Greek goddess Tyche and the Roman goddess Fortuna. Of course, you remember that one of the sons of Jacob, and thus one of the tribes of Israel, was named Gad.

When Leah saw that she had ceased bearing children, she took her servant Zilpah and gave her to Jacob as a wife. Then Leah's servant Zilpah bore Jacob a son. And Leah said, "Good fortune has come!" so she called his name Gad. (Genesis 30:9–11, ESV)

Kind of puts that event in a whole different light.

Meni (Fortune) is mentioned only once in the Bible and not all that often outside it. It's difficult to tell whether Meni was male or female, although some scholars think the name might come from Menītum, an epithet of Ishtar found in a Mesopotamian god-list.

But it doesn't matter. The takeaway is that by the time of King Hezekiah, some 250 years after the construction of the Temple on Zion, at least some Jews had forgotten the source of their blessings and were trusting instead to Gad and Meni—the gods Fortune and Destiny.

———◆———

Jesus, of course, was fully aware of the ongoing war for His holy mountain. For Him, the war was personal.

Many of the key events in the life of Jesus occurred at the Temple Mount. As an infant, Jesus was presented at the Temple in accordance with the Law, where Simeon, a man who had been told he'd live to see the Messiah, and Anna, an 84-year-old prophetess, were led to Jesus by the Holy Spirit. When Jesus was twelve, He stayed behind in the Temple after His parents started back to Nazareth after the Passover celebration in Jerusalem. It was a full day before they realized Jesus was missing, and at least three more before they found Him in the Temple talking with the rabbis.

Early in His ministry, Jesus visited Jerusalem during Passover and drove the moneychangers and animal merchants out of the Temple. Later, probably during the second Passover of His ministry, Jesus healed a lame man at the Pool of Bethesda at the north end of the temple complex. Shortly before the Crucifixion, Jesus drove out the moneychangers a second time, and Matthew records that He healed many lame and blind people who came to Him at the Temple.

Isn't it interesting that even in the building erected by the wicked king Herod, and without the Ark of the Covenant in the temple, Jesus was still consumed with zeal for His Father's house?

And that passion extended beyond the thirty-five acres that make up the Temple Mount. Israel's inheritance was Yahweh, and the land inside the borders He established during the time of Moses and Joshua belonged to Him. That's why Jesus devoted so much of His ministry to healing the sick and casting out demons—which were, remember, the spirits of the Nephilim. He wasn't just restoring people to physical and spiritual health, He was casting the spirits of the giants out of His land, Israel.

Even the Gadarene demoniac who hosted the demons called Legion, speculatively linked to the Amorite "Thousand Gods," Līm, was on Yahweh's turf. Gadara was east of the Sea of Galilee on land allotted by Joshua to the half-tribe of Manasseh.

When we step back and take a fresh look at the events of Jesus' life, many things take on new meaning when they're framed in the context of the war between God and the gods. And, of course, many of the arguments offered by skeptics to explain away the divinity of Jesus are nothing more than PSYOPs by the Fallen to convince modern minds, clouded by the fog of scientism, that Jesus was either a political radical, a social justice warrior, or a misunderstood itinerant preacher—anything but God made flesh.

For example, the Transfiguration. What was the point of all that?

And after six days Jesus took with him Peter and James and John, and led them up a high mountain by themselves. And he was transfigured before them, and his clothes became radiant, intensely white, as no one on earth could bleach them. And there appeared to them Elijah with Moses, and they were talking with Jesus. And Peter said to Jesus, "Rabbi, it is good that we are here. Let us make three tents, one for you and one for Moses and one for Elijah."

For he did not know what to say, for they were terrified. And a cloud overshadowed them, and a voice came out of the cloud, "This is my beloved Son; listen to him." And suddenly, looking around, they no longer saw anyone with them but Jesus only. (Mark 9:2–8, ESV)

First reaction by the modern reader: *Awesome special effects!* But if the purpose of the Transfiguration was to demonstrate His divinity, why did Jesus take only three of His disciples? Wouldn't it have been more productive to bring all twelve? Or why not perform this great visual effect for the thousands of people who had to settle for a miraculous lunch of bread and fish?

Here's why: The intended audience wasn't human.

The clue to the real purpose of the event is the location. Both Mark and Matthew note that the Transfiguration took place on a high mountain. Now, Israel has plenty of mountains, but not many that can be described as high—at least not relative to the rest of the peaks in the land.

Second, note that they climbed a mountain near Caesarea Philippi. That narrows the field. The town was in the northeastern part of the Holy Land, north of the Sea of Galilee and Lake Huleh, in the area we call the Golan Heights. To be precise, Caesarea Philippi sat at the southwestern base of Mount Hermon.

Ah. Bull's eye!

Yes, the very mountain where the Watchers/Titans descended and made a pact to corrupt humanity was where Jesus was transfigured into a being of light before the eyes of Peter, James, and John.

Coincidence?

Not on your life! Jesus knew *exactly* what He was doing. This was a cosmic poke in the eye, a declaration to the Fallen that the Second Power in Heaven had arrived in the flesh. He was declaring that the temporary dominion of the rebellious bene elohim was nearly at an end—that Yahweh's mount of assembly would soon fulfill the promises proclaimed by the prophets. And He did it on the mount of assembly of El, the high god of the Canaanites.

More accurately, it was where the Fallen tried to usurp the name of El Shaddai, the God of Abraham, Isaac, and Jacob.

If we back up a chapter in Matthew and Mark, there is another incident that foreshadowed what happened on Mount Hermon. At the base of the mountain, outside the city of Caesarea Philippi, is a place called Paneas (today called Banias). It's a cave and a spring that's been sacred to the Greek god Pan since the time of Alexander the Great.

Pan was the god of wild, desolate places, shepherds and flocks, music, fields, groves, and wooded glens. Pan was also linked to fertility, which influenced much of the art depicting the god in the classical world. He was often shown pursuing or engaged in physical relations with goddesses, nymphs, women, teen boys called *eromenoi*, and/or goats. (Ecch. Some of those ancient vases and frescoes should have been rated R or NC-17.) As a rustic nature god, Pan normally wasn't worshiped in temples. He preferred outdoor settings, especially those that were like his home in Arcadia, a mountainous region of southern Greece.

The name of the goat-god is the root of the modern English words "panic" and "pandemonium." This comes from a myth in which Pan scared off a group of hostile Titans by letting out a horrible and terrifying screech, thus saving Mount Olympus from the angry giants. Pan was also credited with giving the outnumbered Athenians victory at the famous Battle of Marathon in 490 B.C. by causing the larger Persian force to— you guessed it—panic.

The Grotto of Pan at Paneas has been sacred since ancient times. A gushing spring once flowed from the cave, feeding the marshy area north of Lake Huleh that was the source of the Jordan River. An earthquake years ago shifted something under the mountain, and today the stream seeps quietly from the bedrock below the mouth of the cave.

After the Greeks came to the Levant, Pan gradually replaced an earlier local fertility cult. Scholars have suggested that Aliyan, a minor Canaanite god of fountains, may have been the deity worshiped at Paneas before the

Greeks arrived. This might have been the god called Baal-Hermon, the Lord of Hermon, in Judges 3:3.

Christians in the centuries after Jesus equated Pan with Satan. It doesn't take a lot of imagination to see how Greek depictions of the goat-god influenced later artists, what with the horns, hooves, tail, and all. But those connections were mainly in the minds of imaginative medieval artists, since there is nothing definite in the Bible that describes the appearance of Satan other than a warning that he can appear as an angel of light. However, there are some interesting links that suggest there was more to Pan than a fun-loving, randy nature spirit.

We go back more than 1,400 years from the time of Jesus to the Exodus. Apparently, the Israelites began to worship entities during their forty years in the desert called *se'irim*, or "goat demons."

> And the LORD spoke to Moses, saying, Speak to Aaron and his sons and to all the people of Israel and say to them, This is the thing that the LORD has commanded. **If any one of the house of Israel kills an ox or a lamb or a goat in the camp, or kills it outside the camp, and does not bring it to the entrance of the tent of meeting to offer it as a gift to the LORD in front of the tabernacle of the LORD, bloodguilt shall be imputed to that man.** He has shed blood, and that man shall be cut off from among his people. This is to the end that the people of Israel may bring their sacrifices that they sacrifice in the open field, that they may bring them to the LORD, to the priest at the entrance of the tent of meeting, and sacrifice them as sacrifices of peace offerings to the LORD. And the priest shall throw the blood on the altar of the LORD at the entrance of the tent of meeting and burn the fat for a pleasing aroma to the LORD. **So they shall no more sacrifice their sacrifices to goat demons, after whom they whore.** This shall be a statute forever for them throughout their generations. (Leviticus 17:1–7, ESV, emphasis added)

The *se'irim*, literally "hairy ones," were satyr-like (i.e., Pan-like) beings whom the Israelites began to worship during their wanderings in the desert. The sacrifices away from the tabernacle were consistent with the worship of Pan. It appears that the section of the Law recorded in Leviticus 17, requiring that all sacrifices be brought to the tent of meeting, was specifically to stop the worship of these goat-demons.

How serious were these *se'irim*? In other words, were they real? You could make a case that Yahweh was only forbidding the spiritual practice not because the *se'irim* existed, but simply because sacrificing to these imaginary beings took the place of devotion to Him.

On the other hand, how likely is it that the Holy Spirit gave bad information to His prophets? In his prophecy of a catastrophe about to fall on Babylon, Isaiah writes:

> And Babylon, the glory of kingdoms,
> the splendor and pomp of the Chaldeans,
> will be like Sodom and Gomorrah
> when God overthrew them.
> It will never be inhabited
> or lived in for all generations;
> no Arab will pitch his tent there;
> no shepherds will make their flocks lie down there.
> But wild animals will lie down there,
> and their houses will be full of howling creatures;
> there ostriches will dwell,
> **and there wild goats will dance.**
> (Isaiah 13:19–21, ESV, emphasis added)

The "wild goats" of verse 21 are the *se'irim*, rendered "satyr" in some Bibles, but it's the same word translated "goat demons" in Leviticus 17.

Now, *se'ir* is one of those words where the translation depends on the context. Usually, it simply means "goat" or "kid." But there are four places in the Old Testament where the word clearly refers to a demon or devil. Here's another example: In Isaiah 34, the prophet turns his polemical gift on the land of Edom:

> Thorns shall grow over its strongholds,
> nettles and thistles in its fortresses.
> It shall be the haunt of jackals,
> an abode for ostriches.
> And wild animals shall meet with hyenas;
> **the wild goat shall cry to his fellow;**
> **indeed, there the night bird settles**
> **and finds for herself a resting place.**
> (Isaiah 34:13–14, ESV, emphasis added)

Again, the "wild goat" is based on the Hebrew root *se'ir*. The KJV translates the word as "satyr" as it does in Isaiah 13, but most English translations are similar to the ESV—wild goat, male goat, hairy goat, etc.

Interestingly, there is another demon mentioned in that verse—the "night bird," variously translated in other English bibles as "screech owl," "night creatures," and "the night monster." It comes from the Hebrew word *liyliyth*. That's based on an Akkadian loan word, *lilitu*, which refers to the demoness Lilith.

Even with Yahweh's clear warning against sacrificing to goat demons, about five hundred years later, Jeroboam, after leading the northern tribes into rebellion against Solomon's son, Rehoboam, didn't just set up golden calves at Bethel and Dan for Israel to worship.

> For the Levites left their common lands and their holdings and came to Judah and Jerusalem, because Jeroboam and his sons cast them out from serving as priests of the LORD, and **he appointed his own priests for the high places and for the goat idols and for the calves that he had made.** (2 Chronicles 11:14–15, ESV, emphasis added)

By the way, the passage above debunks the theory that the ten northern tribes were carried off by the Assyrians, leaving only Judah and Benjamin to finish the journey to the end of the Bible alone. The tribe of Levi had been evicted from the northern kingdom by Jeroboam because of their loyalty to Yahweh. (There are other passages in the Bible that contradict the "ten lost tribes" myth, but I digress.)

Now, back to Moses: The book of Leviticus records an interesting requirement for the Day of Atonement. It involved a goat that was driven from the camp into the wilderness for a being called Azazel.

> Aaron shall offer the bull as a sin offering for himself and shall make atonement for himself and for his house. Then he shall take the two goats and set them before the LORD at the entrance of the tent of meeting. And Aaron shall cast lots over the two goats, one lot for the LORD and the other lot for Azazel. And Aaron shall present the goat on which the lot fell for the LORD and use it as a sin offering, **but the goat on which the lot fell for Azazel shall be presented alive before the LORD to make atonement over it, that it may be sent away into the wilderness to Azazel.** (Leviticus 16:6–10, ESV, emphasis added)

Scholars interpret the ritual as complementary rites of atonement, with the sacrificial goat and the second goat who carries the sin of the people away. The priest would lay hands on the goat for Azazel and impart the sins of the people on it before it was led out of the camp and into the wilderness. This is the origin of the term "scapegoat."

Because you've been paying attention, you're remembering right about now that the name Azazel has come up once before in this book. That's because the Book of 1 Enoch names Azazel as one of the leaders of the rebellious Watchers who descended at Mount Hermon. And right there is a link between Azazel and the *se'irim*, the satyr-like deities dancing in the wilderness who just happen to bear a strong resemblance to the Greek god Pan.

Of course, unless you're a coincidence theorist, you see these details as a connected thread, part of an enemy PSYOP to distract the Jews from their devotion to Yahweh. But that's not all.

Greek myths record fascinating connections between Pan and a pair of other deities we've already encountered in this study: In one story, Pan, in his goat-god aspect Aegipan, assisted Zeus in the chief god's epic battle with the chaos god, Typhon. When Typhon turned to attack Aegipan, the goat-god dove into the Nile River, with the parts above the water remaining a goat but the part below the waterline transformed into a fish. Thus, Aegipan became the goat-fish Capricornus, or Capricorn.

Even though the constellation Capricorn is faint, it's been consistently depicted as a hybrid goat-fish since at least the twenty-first century B.C., the time of the last Sumerian kings to rule over all of Mesopotamia. And—here's the kicker—the goat-fish was a well-known symbol of the god of the *abzu*, Enki.

So now we can draw links directly from Enki, god of the abyss, to the goat-demons of the Exodus and the goat-god worshiped at the base of Mount Hermon in the time of Jesus Christ.

And it was there at Caesarea Philippi, right outside the Grotto of Pan, where this exchange took place between Jesus and His hot-headed disciple, Peter:

> Now when Jesus came into the district of Caesarea Philippi, he asked his disciples, "Who do people say that the Son of Man is?" And they said, "Some say John the Baptist, others say Elijah, and others Jeremiah or one of the prophets."
>
> **He said to them, "But who do you say that I am?" Simon Peter replied, "You are the Christ, the Son of the living God."**
>
> And Jesus answered him, "Blessed are you, Simon Bar-Jonah! For flesh and blood has not revealed this to you, but my Father who is in heaven. And I tell you, you are Peter, and **on this rock I will build my church, and the gates of hell shall not prevail against it."** (Matthew 16:13–18, ESV, emphasis added)

Remember that the Grotto of Pan is at the foot of El's mount of assembly, site of the rebellion by the Watchers. It's also in Bashan, gateway to the netherworld. Was it a coincidence that it was there Jesus asked Peter, "Who do you say that I am?"

Seeing as how that's the only question on the entrance exam for heaven, no. Jesus made a statement to the Fallen there, a declaration of his divinity: On *their* rock, He would build his church!

Then He *climbed* their rock, Mount Hermon, and was transfigured into a being of light, His face like the sun and His clothes dazzling white—and the disciples heard Yahweh's voice from heaven.

Peter got it right. And as with many of the other incidents in this book, it took place at a specific location for Yahweh's reasons—in this case, to serve notice to the Fallen that the Messiah had arrived. And the transfiguration that followed on Mount Hermon was a demonstration of authority.

Abode of the gods? Ha! They were powerless to stop Jesus on what was supposed to be their home turf.

Of course, the Enemy was furious. Luke records that Jesus and the disciples found the rest of the disciples in the middle of an uproar when they came down from the mountain.

And when they came to the disciples, they saw a great crowd around them, and scribes arguing with them. And immediately all the crowd, when they saw him, were greatly amazed and ran up to him and greeted him. And he asked them, "What are you arguing about with them?"

And someone from the crowd answered him, "Teacher, I brought my son to you, for he has a spirit that makes him mute. And whenever it seizes him, it throws him down, and he foams and grinds his teeth and becomes rigid. So I asked your disciples to cast it out, and they were not able."

And he answered them, "O faithless generation, how long am I to be with you? How long am I to bear with you? Bring him to me."

And they brought the boy to him. And when the spirit saw him, immediately it convulsed the boy, and he fell on the ground and rolled about, foaming at the mouth.

And Jesus asked his father, "How long has this been happening to him?" And he said, "From childhood. And it has often cast him into fire and into water, to destroy him. But if you can do anything, have compassion on us and help us."

And Jesus said to him, "If you can! All things are possible for one who believes." Immediately the father of the child cried out and said, "I believe; help my unbelief!" And when Jesus saw that a crowd came running together, he rebuked the unclean spirit, saying to it, "You mute and deaf spirit, I command you, come out of him and never enter him again."

And after crying out and convulsing him terribly, it came out, and the boy was like a corpse, so that most of them said, "He is dead." But Jesus took him by the hand and lifted him up, and he arose.

And when he had entered the house, his disciples asked him privately, "Why could we not cast it out?" And he said to them, "This kind cannot be driven out by anything but prayer." (Mark 9:14–29, ESV)

You can understand why that passage generates a lot of questions. What did Jesus mean by, "This kind cannot be driven out by anything but prayer"? Are there ranks within the realm of demons, some that are more powerful than others? What about fallen angels—do they have a hierarchy?

Yes, it appears they do. Peter and Jude reminded their readers that false teachers and false prophets had wormed their way into the body of believers, boldly rejecting authority and blaspheming "the glorious ones." In context, the apostles must have been referring to a class or type of supernatural being.

Jude then mentioned an event not recorded in the Old Testament:

The archangel Michael, "one of the chief princes" (Daniel 10:13), "the great prince who has charge of [Israel]" (Daniel 12:1), did not "presume to pronounce a blasphemous judgment" against Satan while they fought for the body of Moses! (Jude 1:9)

This is slim evidence, sure, but it indicates that there are some unfriendly supernatural entities that even archangels won't try to take down on their own authority.

Not to leave you hanging, but that's a topic for a whole different book. The point in bringing up this episode is to remind you that it wasn't a coincidence that this episode occurred immediately after the Transfiguration on Mount Hermon. While the Fallen knew that they couldn't match the power of Jesus, they *could* make His disciples question the power of the Holy Spirit to work through *them*.

Although we must add, at the risk of being snarky, that doing so right after Jesus set off a supernatural bonfire on the Fallen's counterfeit mount of assembly comes across like a boy shouting insults over his shoulder as he runs away from the kid who just bloodied his nose on the schoolyard.

We need to look at one more event that took place in northern Israel, near Mount Hermon. This is in Luke 10, the chapter immediately after Peter's declaration of faith and the Transfiguration on Mount Hermon. In other words, the time and place of this event was also deliberate.

> **After this the Lord appointed seventy-two others and sent them on ahead of him, two by two, into every town and place where he himself was about to go....**
>
> **The seventy-two returned with joy, saying, "Lord, even the demons are subject to us in your name!" And he said to them, "I saw Satan fall like lightning from heaven.** Behold, I have given you authority to tread on serpents and scorpions, and over all the power of the enemy, and nothing shall hurt you. Nevertheless, do not rejoice in this, that the spirits are subject to you, but rejoice that your names are written in heaven." (Luke 10:1–20, ESV, emphasis added)

Now, some translations say there were seventy disciples, others say seventy-two. The Greek manuscripts of Luke are split and there is no way to know for sure which number Luke wrote on his original draft of the gospel. But either way, given the location and the purpose of the mission, there are good theological reasons to believe the number was not, you know, coincidental.

How many sons of El were in El's assembly on Mount Hermon? That's right, seventy—plus Ba`al and El equals seventy-two.

How many nations did Yahweh create at the Tower of Babel? Seventy—but some translations of the Bible actually name seventy-two.

How many elders of Israel climbed Mount Sinai to meet Yahweh face to face? Seventy—plus Moses and Aaron makes seventy-two.

What was the point of Jesus sending the disciples "to every town and place where he himself was about to go"? It was the start of the church's mission to reclaim the nations from the Fallen, the kickoff event of the Great Commission. Satan, the nachash from the book of Genesis, fell "like lightning from heaven" because he'd lost his legal claim as lord of the dead over those who die in Christ. Why? Because we are guaranteed eternal life through the shedding of His blood.

However you count it, sending out that specific number of disciples wasn't an accident or a coincidence, it was a clear message to the old gods: *Get off my land!*

We mentioned earlier that Ba`al was a thorn in the side of God's chosen people, Israel, through the time of Jesus. We don't think much about Ba`al in the New Testament, but he's there and the link to Old Testament theology is fascinating. It not only brings the ancient Semitic storm-god into the Apostolic Age, it sheds some light on the true identity of this god with many names.

To recap, we showed earlier how Ba`al, which just means "lord," was a god who first appears in history as Iškur, a minor player in the Sumerian pantheon. As time passed, Iškur spread to surrounding cultures in the ancient Near East, known at different places and times as Teššub, Teisheba, Theispas, Tarhun/Tarhunt, Addu/Haddu, Adad/Hadad (the variant most familiar to the Israelites), Set/Seth, Zeus, Jupiter, probably Thor, and possibly even Perun, the Slavic storm-god, and Indra, the Indian storm-god.

By the time of Jesus, Ba`al was still relevant in Jewish religious thought. In fact, the Pharisees tried to accuse Jesus of drawing His supernatural power from Ba`al!

> Then a demon-oppressed man who was blind and mute was brought to him, and he healed him, so that the man spoke and saw. And all the people were amazed, and said, "Can this be the Son of David?" But when the Pharisees heard it, they said, "**It is only by Beelzebul, the prince of demons, that this man casts out demons.**" (Matthew 12:22–24, ESV, emphasis added)

Beelzebul simply means "Ba`al the Prince." And, as you can see above, he was considered the prince of evil spirits by the men who felt threatened by the growing popularity of Jesus. Well, Jesus' reply is enlightening:

> Knowing their thoughts, he said to them, "Every kingdom divided against itself is laid waste, and no city or house divided against itself will stand. **And if Satan casts out Satan, he is divided against himself.** How then will his kingdom stand? And if I cast out demons by Beelzebul, by whom do your sons cast them out? Therefore they will be your judges. But if it is by the Spirit of God that I cast out demons, then the kingdom of God has come upon you. (Matthew 12:25–18, ESV, emphasis added)

See what He did there? Jesus specifically linked the storm-god Ba`al with Satan. Was this a figure of speech? A clever turn of phrase to shut down the Pharisees? No, of course not! If we start substituting what we think Jesus *really* meant for verses in the Bible, we'll have a cult started faster than you can shake a stick.

Consider the implications: By linking Ba`al with Satan, Jesus also linked Satan with two of the most popular pagan gods in modern pop culture—Zeus (remember Liam Neeson in *Clash of the Titans* snarling, "Release the kraken!") and one of the staples of the Marvel superhero universe, Thor.

Are we stretching too far to balance the equation Zeus = Ba`al? No. Not in the least. Look at Jesus' message through John the Revelator to the church of Pergamum:

> And to the angel of the church in Pergamum write: The words of him who has the sharp two-edged sword.
>
> **I know where you dwell, where Satan's throne is. Yet you hold fast my name, and you did not deny my faith even in the days of Antipas my faithful witness, who was killed among you, where Satan dwells.** But I have a few things against you: you have some there who hold the teaching of Balaam, who taught Balak to put a stumbling block before the sons of Israel, so that they might eat food sacrificed to idols and practice sexual immorality. (Revelation 2:12–14, ESV, emphasis added)

Satan's throne? Where Satan dwells? What?

For context, Pergamum (or Pergamon) was a city in what is today far western Turkey. Back in the day, it rivaled Alexandria, Antioch, and Ephesus for culture and wealth. It was strategically situated on land and sea trade routes, with a mountain-top view of the Aegean Sea. Pergamon enjoyed a long period of prosperity from the time of Alexander in the fourth century B.C. until the third century A.D. It was badly damaged by an earthquake in the middle of that century and sacked by the Goths not long after that.

Scholars mainly agree that what Jesus was referring to was the Great

Altar of Pergamon, a massive construction from the Temple of Zeus probably completed in the mid-second century B.C. It's a U-shaped platform
about 116 feet wide by 110 feet deep. The stairway leading up to the
altar is nearly 60 feet wide, and the base of the platform is decorated with
scenes from the Gigantomachy, a battle between the gods of Olympus
and the Giants, children of Gaia (Earth) born from the blood shed by
Uranus (Sky) when he was castrated by their son, Kronos.

And you thought *your* family had issues.

Jesus' reference to the altar from Pergamon is a solid link in the chain
that identifies the great rebel of the Bible through history: Satan is Zeus,
who is Ba`al. And Iškur, the earlier manifestation of the storm-god, as the
lesser twin of the Sumerian god Enki, may be another aspect of the lord of
the *abzu*, the god of the abyss—whom we've linked to the god Pan, worshiped at the cave at the base of Mount Hermon, El's mount of assembly
where the Watchers/Titans began their rebellion.

It's a tangled web, but it's all connected. The supernatural Fallen get to
play with more dimensions than we do.

Would it surprise you to learn that the Great Altar of Pergamon, that
ancient monument to Zeus, was disassembled by German archaeologists
and reconstructed at a museum in Berlin in 1930?

You get the significance. Satan's seat was brought to the capital city of
the Weimar Republic just before the rise of Adolf Hitler and the Nazis.

Coincidence?

———————

Let's turn back to the future capital of God's kingdom on earth. Jerusalem
and the Temple Mount will play a central role in the events to unfold in
the end times.

Our understanding of "cosmic north" could put the war of Gog and
Magog in a new light. Look at Ezekiel 38, where he describes the hordes
of Gog:

The word of the LORD came to me:

"Son of man, set your face toward Gog, of the land of Magog, the chief prince of Meshech and Tubal, and prophesy against him and say, Thus says the Lord GOD: Behold, I am against you, O Gog, chief prince of Meshech and Tubal. And I will turn you about and put hooks into your jaws, and I will bring you out, and all your army, horses and horsemen, all of them clothed in full armor, a great host, all of them with buckler and shield, wielding swords. Persia, Cush, and Put are with them, all of them with shield and helmet; Gomer and all his hordes; **Beth-togarmah from the uttermost parts of the north with all his hordes**—many peoples are with you....

"Therefore, son of man, prophesy, and say to Gog, Thus says the Lord GOD: On that day when my people Israel are dwelling securely, will you not know it? **You will come from your place out of the uttermost parts of the north**, you and many peoples with you, all of them riding on horses, a great host, a mighty army." (Ezekiel 38:1–6, 14–15, ESV, emphasis added)

The Hebrew words translated "uttermost parts of the north," *yerekah tsaphon*, appear in only three places in the Old Testament. If you've read from the beginning of the book to this point, you've encountered all three: Isaiah 14, referring to Ba`al's mount of assembly; Psalm 48, comparing Yahweh's mount of assembly to Ba`al's; and Ezekiel 38, apparently referring to the origin point for the army led by Gog of Magog.

But what if Ezekiel was referring to the *spiritual* origin of the army of Gog rather than a point on a map? Consider how differently we'd look at this passage if Ezekiel 38 was translated this way:

And I will turn you about and put hooks into your jaws, and I will bring you out, and all your army, horses and horsemen, all of them clothed in full armor, a great host, all of them with buckler and shield, wielding swords. Persia, Cush, and Put are with them, all of them with shield and helmet; Gomer and all his hordes; **Beth-togarmah from the utmost heights of Mount Zaphon** with all his hordes—many peoples are with you....

Therefore, son of man, prophesy, and say to Gog, Thus says the Lord GOD: On that day when my people Israel are dwelling securely, will you not know it? **You will come from your place on the remote slopes of Zaphon**, you and many peoples with you, all of them riding on horses, a great host, a mighty army. (Ezekiel 38:4–6, 14–15, ESV, words in bold substituted by this author)

Admittedly, the context of a passage dictates its correct translation. While several English language Bibles see Mount Zaphon rather than "parts of the north" in Isaiah 14, there are none that render Ezekiel 38 this way. So this author may be alone on a hill on this one.

We don't presume to know more than the scholars who've translated the original Hebrew. Still, it's intriguing that the same phrase used to describe the divine rebel's mount of assembly in Isaiah 14, the holy mountain of Ba`al, is used by Ezekiel to identify the point of origin of the army led by Gog of Magog. And since Mount Zaphon, Jebel al-Aqra, is in Turkey, "the sides of the north" is in the general area of the House of Togarmah and several of the other nations named by the prophet.

Given recent geopolitical developments in that region, especially along the border between Turkey and Syria, it is especially interesting.

Speculating a bit here, Armenians consider Togarmah the father of the patriarch of their nation, Hayk. In Armenian legend, Hayk killed the evil king of Babylon, Bel, who is identified with Nimrod. According to the story, from the time of their ancestors Noah and Japheth, Hayk and his people had migrated south toward the warmer lands around Babylon. However, an evil giant, Bel, ruled Babylon with an iron fist and tried to enslave Hayk's people. Instead, as soon as his soon Aramaneak was born, Hayk led his people back north to the land of Ararat (possibly Aratta).

Hayk and his men found Bel and his army holding a mountain pass southeast of Lake Van, a large lake in what is today far eastern Turkey. In the battle that followed, Hayk managed to kill Bel with a nearly impossible shot from his longbow (on August 11, 2496 B.C., per tradition). A giant statue of Hayk drawing his bow is proudly displayed in Armenia's capital, Yerevan, to this day.

While this is just a legend, it does echo some elements of documented history. After the Flood, descendants of Noah did migrate south and east from the mountains of Ararat. A king or series of kings from the Sumerian city of Uruk, the capital of Nimrod's domain in the region later called Babylon, spread the influence of Uruk at least partly by force, documented by the ruins of Hamoukar in far northeastern Syria. For all we know, the legend of Hayk preserves a memory of the conflict between the prehistoric king of Uruk, Enmerkar, and the lord of Aratta.

Returning to the realm of what we can know for sure, wherever Magog and the hordes of its coalition come from, we know where they're going: Zion. That is where the mother of all battles will be fought, Armageddon.

The book of Revelation tells us that the armies of the world will gather "at the place that in Hebrew is called Armageddon" (Revelation 16:16). Many scholars believe the word, which appears only once in the Bible, is derived from the Hebrew words *har m giddô*, or Mount of Megiddo.

This would seem to make sense. As we mentioned earlier, Megiddo was the site of several famous battles in history because of its strategic value at the head of a pass through the Carmel Ridge, overlooking the Valley of Jezreel. Besides the two battles we named earlier, King Josiah of Judah was mortally wounded at the Battle of Megiddo in 609 B.C. against the Egyptian army led by Pharaoh Necho II.

However, there is a fundamental problem with placing the Battle of Armageddon at Megiddo: There is no mountain at Megiddo. It's a valley.

Here's the thing: In Greek translations of Hebrew, it is common for the letter *gamma* to represent the Hebrew letter *ayin*, so the original title *har mo'ed*—"mount of assembly"—was transformed into *har məgiddô*.

Reading Armageddon as *har mo'ed*, the mount of assembly, makes a lot more sense than trying to place a mountain in a valley. Most of the apocalyptic prophecy in the Old Testament centers on Jerusalem and

Mount Zion—for example, Joel 3 and Zechariah 14. So it's far more likely that the word "Armageddon" is based on misunderstanding how Hebrew is transliterated into Greek than on a prophecy of a battle at a nonexistent mountain.

The battle that's become a byword for the most destructive event humans can imagine, the end of all things, will be fought at Jerusalem for control of the Temple Mount—Mount Zion.

SUMMING UP

Supernatural warfare in the Bible isn't limited to the Old Testament. When you consider that demons are the spirits of the Nephilim who died in the flood of Noah, this supernatural war is a long series of connected battles, all part of the same conflict that's raged since Eden.

Many of the events of Jesus' life take on new significance in the light of the divine council paradigm. Jesus battled demons to finish clearing the land Yahweh allotted to His people, Israel. The Transfiguration on Mount Hermon was an unmistakable message to the Fallen that the prophesied Messiah had arrived, and sending out the seventy (or seventy-two) disciples emphasized the point that the bene elohim who'd been allotted to the nations after Babel were nearing the end of their time. Jesus even declared, in front of the Grotto of Pan at the foot of Mount Hermon, which some believed was literally the entrance to the netherworld, that He'd build His church upon their rock and the gates of hell would not prevail against it!

No wonder the Enemy spends so much effort on PSYOPs to discredit the literal life, death, and resurrection of Jesus. The eyewitness accounts establish Jesus as historical, and *His* testimony confirms that the Old Testament is reliable. Sadly, neo-atheists are content on the wrong side of history. And too many of us in the church aren't strong enough in apologetics to punch holes in their wet paper bag.

Please—don't let your children or grandchildren fall victim to history deniers. The consequences are eternal.

8

JABAL AL-NOUR

If you've been counting, you know that we have featured seven cosmic mountains thus far in the book. If we were interested in stopping at a nice biblical number to justify a catchy title, something like *The Seven Mountains of the Supernatural War*, we'd stop at Zion and wrap things up. But this book would be incomplete without bringing up mountain number eight.

This mountain has had a tremendous and incredibly destructive impact on the world. It's responsible for what statisticians say will be the world's largest religion by about 2070 unless something drastic happens. In that regard, we must give credit where it's due—this mountain was, without question, the site of the most successful supernatural PSYOP in history.

The Arabic name of this peak means "Mountain of Enlightenment." Jabal al-Nour is near Mecca in western Saudi Arabia. It's the site of the cave where Muhammad was visited by an angel calling itself Jibril, or Gabriel.

The message was not from God. Yahweh does not contradict Himself, but the Quran clearly does. Muhammad saw something, of that we have no doubt. But it was *not* a messenger from Yahweh.

227

We'll deal with the specific beliefs of Islam and its major sects in the next chapter. Islam has a role to play in the future conflict for God's holy mountain, but sadly for its adherents, the part the Fallen wants them to play is very simple—to die.

Allah is the name used for the god of Islam, but it's also the name used for the God of the Bible by Arab Christians. This is somewhat confusing, which is, of course, not coincidental. Just as El was the name of the chief god of the Canaanites *and* a generic name for God in Hebrew, "Allah" is another PSYOP by the Fallen to muddy the water.

Many Christians assume that Muslims worship the same god they do, and so they aren't surprised to learn that Arab Christians apply the same proper name to the God of the Bible. This is the same mistake as assuming a pagan Amorite in the fifteenth century B.C. who prayed to El was calling on the God of Abraham, Isaac, and Jacob. Muslims, who believe that Allah has no son, have put their faith in something they *think* is the God of the Book (the Bible). How can that be, when there are fundamental differences between Allah and Yahweh?

Some Christian scholars draw a link between Allah and older deities worshiped in Arabia prior to Muhammad. This isn't universally accepted, but it doesn't have to be to make a case for the god of Islam being something other than Yahweh.

Etymologically, the explanation that seems to make the most sense, and the one accepted by most scholars, suggests that Allah is a contract of *al-l h*, "the god." That's like the old Semitic use of El, a proper name that grew out of the generic Proto-Semitic word *'il-*, meaning "deity" or "god." Variant forms show up in Akkadian (*ilu*), Ugaritic (*il*), and Hebrew (*eloah*, the singular form of *elohim*).

Still, differences in the fundamental characteristics of Allah and Yahweh force us to conclude that one of two things must be true: Either, 1) Allah and Yahweh are one and the same, and the Bible doesn't accurately record the way Yahweh revealed Himself to the prophets and apostles; or, 2) some *thing* lied to Muhammad in that cave on Jabal al-Nour, and Allah is not the God of the Bible.

Given that we have Old Testament texts that can be reliably dated to at least two centuries before Jesus (the Septuagint translation from Hebrew into Greek), and most scholars would accept earlier dates for the authorship of those books, and that the books of the New Testament are the best attested documents from the classical period, there is compelling evidence that the Bible has not been corrupted or substantially changed since it was written by the apostles and prophets. While there are minor differences between some texts, the sheer number of manuscripts and documents that quote scripture, like letters from the early church fathers, provide plenty of material to cross-check the books of the Bible and confirm its reliability.

In contrast, the Quran was compiled into its final form about twenty years after the death of Muhammad from several competing versions. By the time of the third caliph, Uthman ibn Affan, Islam had spread from Arabia into Iraq, Syria, Egypt, and Iran, cultures that were different from one another and from the origin point of the faith. Uthman was reportedly motivated to take on the project by disputes between some of the new followers over the correct way to pray. So, to prevent Muslims from fighting over the book, texts were collected from around the caliphate and compiled into an "official" version, and variant copies were destroyed.

Even Muslim scholars—mainly Shia, who believe that Muhammad's cousin and son-in-law was the rightful heir to the prophet's growing empire—admit that Uthman corrupted the Islamic holy book by his action.

There is more to the story, of course. Scholars spend entire careers studying the Quran just as others do the Bible. But the fundamental difference between the Quran and the Bible is that there is no comparison between the two when it comes to the quantity and quality of the source material.

We take no joy in saying so. Literally billions of humans will pay the ultimate price for the supernatural deception worked on a charismatic Arab trader fourteen centuries ago. And billions more will be destroyed in the cataclysmic final battle for the holy mountain of Yahweh. The blood-thirsty gods are preparing for war.

———

Here is where another brief review of history may be helpful. We'll get a little speculative, and because we're dealing with the spirit realm there is only so much we can say with one hundred percent confidence.

What happened to the Amorites? Scholars have been exploring that question ever since their existence was confirmed by sources outside the Bible in the nineteenth century. They were pushed out of Mesopotamia by the Kassites, who took Babylon in the sixteenth century B.C., and native rulers near the Persian Gulf called the Sealand Dynasty. We've already discussed how the Amorite Hyksos rulers of Lower Egypt were evicted by native Eygptians around the same time.

They Amorites disappear from history around the time of the Sea Peoples invasion of the Levant and Egypt in the late thirteenth and early twelfth centuries B.C. This coincides with what scholars call the Late Bronze Age collapse. Over a period of about fifty years, between about 1200 and 1150 B.C., the major states of the eastern Mediterranean from Mycenaean Greece and Hatti to the Amorite and Canaanite states of the Levant were destroyed.

This was the event that ended the kingdom of Ugarit during the reign of the ill-fated Ammurapi we mentioned in an earlier chapter. A small kingdom southeast of Ugarit called Amurru, named for the Amorites, was also wiped out around this time.

This was the period of the later Judges, a time that set the stage for the emergence of the kingdom of Israel. Barak and Deborah's destruction of Hazor, which some scholars think may have belonged to the kingdom of Amurru, may be dated to this time.

The bottom line is that only Egypt and a few small kingdoms in northern Syria appear to have survived this wave of destruction. The Philistines settled in what we call today the Gaza Strip, and they may have been part of the Sea Peoples coalition. Emerging from the ashes of this turmoil alongside the kingdom of Israel were several Aramean kingdoms, with Damascus and Hamath being the largest.

Because they came out of the same area formerly associated with the Amorites, it's easy to assume that the Aramean people were just descendants of the Amorites who carried on as best they could after the disasters of the Late Bronze Age collapse. But we can't make that assumption any more than we could assume that the Israelites were just Canaanites with a repackaged religion because of where David came to power.

By analyzing texts that have been found from the Neo-Assyrian, Neo-Babylonian, and later Egyptian kingdoms, we can piece together a history that seems solid. While the name "Amorite" fell out of use after the conquest of Canaan, there are a couple of tribal names that bridge the period between the disappearance of the Amorites and the emergence of the Arameans. Specifically, the Aḫlamū, sometimes called Aḫlamū-Aramayū (Aḫlamū-Arameans) by the Assyrians, and our old friends the Suteans.

Together, these two tribal names appear to be used, at least sometimes, interchangeably with both Amorite and Aramean nomads, mainly in the steppes of Syria and the Transjordan. After the conquests of the Assyrian empire through the seventh century B.C., even the Arameans fade from history, although the Aramaic language, because it was adopted as the *lingua franca* by Assyria, became the language of trade and cultural exchange in the Near East from about 600 B.C. to around the time of Jesus before giving way to Greek.

In short, scholars can document a "geographical, historical and linguistic continuity" between the Amorites of the Old Babylonian period and the Arameans of the Middle Babylonian era.[36]

Recent history in the old Amorite homeland, which includes Iraq, Syria, Jordan, and Lebanon, points to a troubled future for the Holy Land. We mentioned earlier the existence of territorial spirits, the main biblical example being the Prince of Persia that tied up the messenger to Daniel for several weeks. We can't know for certain, but it's worth mentioning because, as you've noticed by now, this author doesn't put much stock in coincidence, especially where it concerns the Bible and the ongoing spiritual war.

Here are the relevant questions: Can it be just a weird coincidence that the hot spots in the Middle East, especially since the emergence of the Islamic State since the summer of 2014, are a match for the areas defined as the homeland of the Amorites four thousand years ago? Is it just a cosmic accident that the land of the Amorites, especially central and northern Syria, are far more important in the eschatology of Islam than Arabia, the land of Islam's birth?

Consider the otherwise inexplicable atrocities in the Islamic State's capital, Raqqa, which lies on the west bank of the Euphrates close by Jebel Bishri, the ancient mountain of the Amorites; or the destruction in Aleppo, called the City of Hadad when it was ruled by Amorite kings in the days of Abraham, Isaac, and Jacob. The prophecies that guide the leadership of the Islamic State focus on lands where the old gods of the Amorites once reigned supreme.

In Iraq, where ISIS is hard pressed at this writing by Iraqi, Kurdish, Turkish, and American forces in the city of Mosul, analysts at the Combating Terrorism Center at the U.S. Military Training Academy at West Point suggest that ISIS has already prepared a fallback plan in Diyala province if they can't hang on in Mosul. The analysts feel the terrain and the demographic makeup of Diyala is especially well suited for an ongoing insurgency.[37] That's the area northeast of Baghdad along the Diyala River toward the Hamrin mountain range, where more than four thousand years ago the doomed Sumerian kings of Ur built their futile Amorite-wall-which-keeps-the-Tidnum-at-bay.

All of this begs the question: What foul spirit is at work in MAR.TU[ki], the ancient land of the Amorites?

———

Let's dig a little deeper. And, if you'll indulge us, we'll get a little speculative. As we've noted, there is only so much we can know for sure about the spirits opposed to God. They lie, and our perception into that realm is limited. But let's look at what's available to us and see if we can draw some tentative conclusions.

Most Bible scholars place two of the nations in Ezekiel's prophecy of the Gog-Magog war, Sheba and Dedan, in Arabia. Sheba was the father of the Sabeans, who founded a kingdom in southwest Arabia, modern-day Yemen.

Dedan settled along the coast of the Red Sea in western Arabia, in the area called the Hejaz. Dedan was an important oasis along the caravan route between Sheba and Babylon. Because of the brutal desert that covers the interior of Arabia, the route traveled north through Edom, just southeast of Judah. Dedan eventually grew into an independent kingdom around the time of the prophets Ezekiel, Jeremiah, and Daniel, in the seventh and sixth centuries B.C. The last king of Babylon, Nabonidus, spent most of his reign living at the important Dedanite oasis of Tayma while entrusting Babylon to his son, Belshazzar—he of the *mene, mene, tekel, upharsin* incident.

Sheba and Dedan were sons of Raamah, a son of Cush, the son of Ham. This makes Sheba and Dedan nephews of Nimrod, who we believe was the Sumerian king Enmerkar, would-be builder of the abode of the gods, the tower of Babel at Eridu. (Sheba and Dedan are also mentioned in Genesis 25 as grandsons of Abraham and his concubine Keturah, through their son Jokshan. Geographically, though, it still places them among the tribes of Arabia.)

Nobody is sure why Nabonidus spent a decade living in the desert, but the best guess is that he was seeking prophecy and guidance from his preferred deity, the moon-god Sîn, who the king elevated to the top spot in the pantheon.

I accomplished the command of Sîn, king of the gods, lord of lords, dwelling in the heavens, who, in comparison of the gods in heaven, his name is surpassing: (also) of Šamaš *(sun-god)*, who is his brightest (peer), of Nusku *(fire-god)*, Istar *(Ishtar/Inanna)*,

Adda *(Addu/Hadad, the storm-god Ba`al)*, Nergal *(Resheph/Apollo)*, (those) who accomplish the command of Nannar *(Sumerian name for Sîn)* their surpasser. —*Harran inscription H2, A & B, Col. III*[38]

Note that Marduk is missing from that list of deities. Nabonidus probably made an enemy of the established Marduk priesthood in Babylon, possibly creating a religious fifth column that contributed to the ease with which Babylon fell to the Persians.

It's believed that Nabonidus' mother Addagoppe (Adad-guppi— notice the theophoric element Adad) was a priestess of Sîn from Harran in northern Mesopotamia. Remember, Harran was the moon-god's cult center that was so important to the Amorite Binū Yamina tribes in Abraham's day, 1,400 years earlier. So Nabonidus may have been of old Aramean/Amorite stock, and for some reason—call it infernal revelation, if you will—he was compelled to revive the flagging cult of Sîn and transplant it to the Arabian desert.

Nabonidus, the last king of Babylon, shown worshipping the moon-god Sîn, the sun-god Šamaš, and Ishtar, represented by Venus. (Credit: Jona lendering (Own work), CC BY 3.0, https://commons. wikimedia.org/w/index. php?curid=2629946)

The moon-god has a long history in that part of the world. Not only do we have the evidence of the cult centers of Ur, Harran, and Jericho that date back to the thirdmillennium B.C., but remember the account of Gideon's victory over the Midianites, another people who lived in northwestern Arabia. Gideon's huge haul of gold included the crescent ornaments from the Midianites' camels, which presumably honored the moon-god. Symbols depicting a heavenly triad of Sîn, Šamaš, and Ishtar— moon, sun, and Venus—are common from the mid-second millennium B.C. at least through the time of Nabonidus, who is depicted venerating the three deities on stela that have survived.

While the star and crescent symbol prominent in the Islamic world today only came into use after the Ottomans took Constantinople in 1453, it's not too much of a reach to suggest that the ancient cult of the moon-god in the Near East and Arabia has been carried into the modern era by the descendants of the Amorites, Arameans, and their Arab neighbors to the south, especially with the boost given to the moon-god's cult in Arabia by the king of Babylon during the last days of that empire.

Why is this of interest? While Sheba (Yemen) seems almost geographically irrelevant to end-times prophecy (except maybe as a flashpoint for a wider Sunni-Shia war), Dedan, as we showed earlier, bears the same name as the ancient Amorite tribe Didanu/Tidanu, from which the Greeks derived the name of the Titans. And the area settled by the Dedanites, the Hejaz, includes the two holiest sites of Islam, Mecca and Medina.

While Muslim scholars would argue the connection, the symbolic link, at least, is obvious. Joel Richardson does an excellent job of establishing the pagan roots of Allah in his new book, *Mystery Babylon*. The historical link between Nabonidus, the moon-worshiping king of Babylon, and the region of Islam's holiest sites may be coincidental, but you know our feeling about coincidence theories.

Now, please understand: We're not suggesting a physical connection between the Watchers/Titans and the Arab tribes that spread Islam across the world. In other words, and we truly hope this doesn't disappoint you, we do not see a role in the end times for ISIS Nephilim.[39] Nevertheless, it's fascinating that history has once again provided a *spiritual* link between the past and the future where we never expected to find one.

And in Revelation 18, Dedan may be more central to events than we suspect.

SUMMING UP

The greatest PSYOP the Fallen have worked on humanity over the last six thousand years of recorded history was pulled at Jabal al-Nour near Mecca. An entity calling itself Jibril, the angel Gabriel, set a successful businessman named Muhammad on a new course. The corrupted version of the Bible that Muhammad preached eventually turned most of the Near East and north Africa to his false god within a century of his death. Syria was the first province of the Byzantine empire to fall to the caliphate. Mesopotamia was taken from the Persians soon after. The Ottoman Empire was arguably the longest-lived empire in history at the time of its dissolution in 1922. At its current rate of growth, Islam will become the world's largest religion by 2070.

These basic facts, combined with unending violence in the old Amorite homeland specifically intended to trigger the end times, leads to uncomfortable questions: How are the spirits of the Nephilim/Rephaim, the deadly offspring of the Watchers/Titans, influencing events in the world today?

And what about the gods who still walk the earth, those not chained in Tartarus like the Watchers? What influence do they have on the unfolding of end times prophecy?

In short: What's their next move?

9

WHAT THINGS MAY COME

WE SHIFT NOW from an analysis of history to a look at what things may come. The Bible tells us a lot about a coming Day of the Lord, when Yahweh finally executes judgment on a fallen world. A day is also coming when the rebellious *elohim* get their due—the day the gods die like men.

Bible prophecy tells us what the future holds. But before we get into the meat of this section, let me make one observation.

The fastest way to get otherwise mild-mannered Christians to fight is to ask, "Pre-, mid-, or post?" That refers, of course, to the timing of the Rapture, and whether it comes before, during, or after a seven-year period of hell on earth called the Great Tribulation.

It's obvious that we Christians have forgotten that even the apostles didn't fully understand the prophecies of Christ's first coming until after the fact. See Acts chapter 1: Forty days after the Resurrection, they were *still* asking him, "Lord, will you at this time restore the kingdom to Israel?"

They were so convinced that the prophecies of Messiah pointed to a geopolitical savior that they missed the actual goal of Christ's mission. Sadly, of course, so did the Jewish religious leaders of Jesus' day, which is why Jews even today are still waiting for the first appearance of the *mashiach*.

When you read from Acts chapter 1 into the second chapter, you get to Pentecost. Holy Spirit descends, disciples speak in tongues, boom. *Then* they got it.

The tongues were not personal prayer languages, by the way. People from all over the world could suddenly understand these Spirit-filled men from Galilee. Understand this point: They heard the gospel *in their native tongues*. That was a 180-degree reversal of the confusion of languages at the Tower Babel.

Coincidence?

But here's the point of this introduction: God, being a good military commander, doesn't share His strategy with the Enemy. He's called the Lord of Armies for a reason (that's what "hosts" means, in case you've wondered). Yahweh deliberately obscured the prophecies of Messiah's first coming so that the Fallen wouldn't understand why Jesus was here.

This is not speculation. This is an important point the Church would do well to remember when we study and discuss eschatology. Paul explained it to the church at Corinth:

> Yet among the mature we do impart wisdom, although it is not a wisdom of this age or of the rulers of this age, who are doomed to pass away. (1 Corinthians 2:6, ESV)

The Greek word translated "rulers" in this passage is *archon*. It can mean humans or supernatural beings. It's the same word John used to record Jesus' description of "the ruler of this world" who was about to be cast out—Satan. And note Paul's reminder that the death of the gods is a dead-nuts lock. They're doomed.

Paul's message to the church at Corinth was simple and profound: "The supernatural principalities and powers didn't understand what I'm about to share with you."

But we impart a secret and hidden wisdom of God, which God decreed before the ages for our glory. **None of the rulers of this age understood this, for if they had, they would not have crucified the Lord of glory.** (1 Corinthians 2:7–8, ESV, emphasis added)

Please understand the significance of that passage. The principalities and powers thought they were putting a spike in God's plan when they drove the nails into Jesus. Paul tells us that if they'd known that they were actually *fulfilling* Jesus' mission, they wouldn't have crucified Him!

That was an epic supernatural judo move, taking the Enemy's most intense desire—to strike a blow at Yahweh Himself—and using it against them.

The prophecies of the Messiah's first coming were so well obscured that the supernatural entities we contend with were fooled. The apostles who learned directly from Jesus were also clueless. It is highly probable that God has likewise arranged the prophecies of the Messiah's Second Coming to once again hide His specific strategy from the Enemy.

Remember, Jesus warned us about a deception that will lead astray even the elect, if it were possible. Don't you think prophecy will play a role in that deception? And wouldn't an intelligent enemy use it as a key weapon to lure people into believing a lie?

In other words, and with all due respect to the many learned men and women who have studied eschatology for years, *we do not fully understand the prophecies of Christ's return.* And we won't understand them until, like the apostles after Pentecost, we can look back at the fulfilled prophecies with 20/20 hindsight.

So please—let's be gracious to one another about differences of interpretation. If end-times prophecies were absolutely clear, we'd all agree on what they mean.

And the Enemy could fabricate an absolutely convincing lie.

That said, there are some basic points of end-times prophecy we must understand. Missing them will lead us into dangerous territory. And most of us are stumbling toward the Apocalypse like blind men in a minefield.

While Christianity is the largest religion on earth with an estimated 2.2 billion believers, it represents less than a third of the world's population. Muslims number about 1.6 billion, or 22 percent, of the global population. (Jews, surprisingly, don't even make the top ten, placing eleventh with about 14 million adherents, only 0.22 percent of the world's population.)[40]

If you're reading this book, you may be surprised to learn that Protestant Christians, especially those with a premillennial, pre-tribulation view of end-times prophecy, are a minority of the world's Christians. Roman Catholic and Orthodox Christians, who are generally amillennial (rejecting the belief that Jesus will have a literal thousand-year reign on earth), comprise 62 percent of global Christianity.[41] Among Protestants, it's safe to say that most of the mainline denominations do not teach, much less emphasize, end-times prophecy.

Making matters worse, biblical illiteracy in America, even among Christians who describe themselves as "born again," is rampant. A 2009 study by the Barna Group revealed that only 9 percent of American adults possess a biblical worldview. But it's not much better among born-again believers: Only 19 percent of people who call themselves born again hold a biblical worldview.[42]

It's no surprise, then, that most American Christians know very little about Bible prophecy, and what we do know might have been picked up from movies or novels.

What's truly frightening about those numbers is a 2013 survey, widely reported by conservative media, trumpeted the finding that two in five American adults believe they're living in the end times. Among Evangelicals, the proportion rose to three out of four.

But if they don't hold a biblical worldview—in other words, if they don't know enough Bible to believe some very basic tenets of Christian doctrine[43]—how in the world can they know what the end times are supposed to look like?

Therein lies the seed of a great deception. It could be the greatest PSYOP in history, one that lures millions of Jews and Christians into welcoming the Antichrist when he appears on the world stage. It is the belief of this author that the rebellious gods have set in motion a brilliant deception that will convince many that the Antichrist is, in fact, the Messiah.

The event that triggers the end of history will take place around an uneven rocky plateau thirty-three miles from the eastern shore of the Mediterranean Sea. The faith of Abraham, the dynasty of David, and the divinity of Jesus were established there. At the heart of this plateau is the hill on which Yahweh, the Creator of the universe, established His "mount of assembly." The temples of Solomon and Zerubbabel were built there, and a prophesied Third Temple will someday occupy that place. The eternal significance of the site was accurately described by the title of David Flynn's 2008 book, *Temple at the Center of Time*.

While setting dates for the fulfillment of end-times prophecy is foolish, since no man knows the day or the hour of Christ's return, a day is fast approaching when Jerusalem's Temple Mount will be at the center of the most important event in human history—the final conflict, a supernatural showdown between Good and Evil at the place the apostle John called Armageddon.

Americans have a difficult time understanding the intensity of the emotions behind the conflict over the Temple Mount. Nationalism, religious fervor, and racism are focused like a laser on an area of only about thirty-five acres. The irrationality spawned by these emotions—for example, the grand mufti of Jerusalem recently declared that the Al-Aqsa Mosque has been on the Temple Mount "since the creation of the world"—suggests that their origin is supernatural.

This speck of land, roughly 0.00000003 percent of the earth's land mass, is the focus of constant confrontation between Muslims, Jews, and Christians. The passions and aspirations that swirl about the Temple Mount will play an ever-increasing role in global geopolitics as the world draws closer to a battle so terrifying that its name is a byword for any type of catastrophe.

The Temple Mount is the holiest site in Judaism, the place where Solomon built the Temple to Yahweh. The Second Temple was constructed there by Zerubbabel, after the decree of the Persian emperor Cyrus, between 538 and 516 B.C. It was rebuilt and expanded by Herod the Great beginning in 19 B.C.

The Temple is at the core of Jewish identity, at least for conservative and orthodox Jews. The Holy of Holies, the inner sanctuary within the tabernacle, was the most sacred site in Judaism, screened from the outer sanctuary by the veil of the covering. The Holy of Holies was the home of the Ark of the Covenant, or Ark of the Testimony, a gold-clad wooden chest that contained the stone tablets on which Yahweh had written the Ten Commandments, Aaron's rod, and a pot of manna. Only the high priest could enter the Holy of Holies and approach the Ark, and then only once a year on the Day of Atonement (Yom Kippur).

Contrary to its description in the blockbuster film *Raiders of the Lost Ark*, it was not "a transmitter…a radio for speaking to God." The Ark of the Covenant was far more important than that—it was literally Yahweh's seat when He appeared among His people (see Exodus 25:22, Numbers 7:89, and 1 Samuel 4:4).

However, both the Ark of the Testimony and the Temple have seemingly been lost to the ages. After the Jewish revolt in A.D. 70, Roman soldiers destroyed and looted Jerusalem, leaving most of Herod's Temple in rubble. The menorah was depicted among the spoils from the Temple on the Arch of Titus in Rome, commemorating his victory over the Jewish rebels.

The Ark disappeared long before the first century A.D. The last sure mention of it in the Bible is 2 Chronicles 35:3, where King Josiah of Judah, who reigned from 640 to 609 B.C., ordered that the Ark be returned to the Temple. One of his predecessors had apparently removed it, for some reason. But after that, for all intents and purposes, the Ark vanished. Jeremiah prophesied a time when the Ark will no longer be discussed or used, and indeed, it was never in the Temple from the time of Zerubbabel until the Temple's destruction. A raised area on the floor of the Holy of Holies indicated where the Ark should have been.

And yet we're entering a period of history in which the Ark (possibly) and the Temple (definitely) will play central roles. A spiritual conflict is brewing that will pit Christians, Jews, and Muslims against one another in a brutal struggle for control of the Temple Mount.

Some want control of the mount for political reasons. Others may believe they can literally trigger the Apocalypse. The Temple Mount and the Ark of the Covenant are right at the center of the oncoming storm.

Let's examine the prophetic expectations of the major players in the coming battle.

———◆———

This author is admittedly not a scholar of the Bible and even less so of the teachings of Jews and Muslims. And we should remember that Islam and Judaism are not monolithic religions. Both of those faiths are subdivided into sects that hold differing and often contradictory beliefs, not unlike us Christians. So please bear in mind that the summary below is not authoritative or exhaustive. Not everyone who identifies as a Muslim will hold all, or any, of these beliefs.

This analysis is meant to demonstrate that the rebellious gods—the principalities, powers, thrones, and dominions referred to by the apostle Paul—have set in motion a brilliant end-times PSYOP. And it appears that Muslims will play the most tragic role of all the world's major religions at the end.

Muslim eschatology is dominated by two prophesied figures. The Mahdi, or "rightly-guided one," is similar to the Jewish *mashiach*, a mortal man who plays a central role in defeating the enemies of Allah. His nemesis is the Dajjal, the Islamic Antichrist figure.

Contrary to the publicly expressed opinions of Western political leaders and progressive pundits, groups such as Jabhat Fateh al-Sham (the former Jabhat al-Nusra) and ISIS do not represent the lunatic fringe of Islamic thought. Their expectation of the Mahdi's imminent arrival is shared by most of the world's Sunni Muslims. In contrast to American Christians, some 80 percent of whom do not expect the literal return of

Jesus anytime soon, upwards of three-quarters of Muslims in the Middle East and South Asia expect to see the Mahdi before they die. Therein lies the power behind ISIS—Mahdism.

> Looking at specific countries, the highest percentage of the population expecting the Mahdi's near-term appearance is found in Afghanistan (83 percent), followed by Iraq (72 percent), Turkey (68 percent) and Tunisia (67 percent). Sixty percent of Pakistanis, 51 percent of Moroccans, 46 percent of Palestinians and 40 percent of Egyptians are looking for the Mahdi in their lifetimes. The conventional wisdom in recent decades among many journalists, and not a few area "experts," has been that Mahdism is an eccentric outlier belief held mainly by (Twelver) Shi`is and the uneducated on the fringes of the Sunni world. This Pew data, among other things, shows the intellectual vacuity of such biases. The average for the 23 countries Pew surveyed on this issue of Mahdism comes out to 42 percent, and extrapolating from that to the entire Muslim world means there are over 670 million Muslims who believe the Mahdi will return here in the first half of the twenty-first century.
>
> What does this Pew information on Mahdism mean? First and foremost, Mahdism must be taken seriously as an intellectual, sociological and even political strain within the entire Islamic world – not dismissed as archaic, mystical nonsense.[44]

This is what we meant in the introduction of the book when we wrote that the Islamic State hasn't hijacked Islam, it is a *purer form* of Islam. And it has a hands-on approach to hot-wiring the Apocalypse.

While many Muslims do not support the methods and/or aims of ISIS, "Islamic history is rife with violent jihads led by self-styled Muslim messiahs and waged by their followers."[45] ISIS caliph Abu Bakr al-Baghdadi and the Islamic State certainly fit the description. And if only 1 percent of the world's Muslims rally to its cause, nearly 7 million jihadists could be available to serve the emerging caliphate.

In the near term, the danger posed by ISIS to the Christian communities of Syria and Iraq is shared by other Muslims in the area. Islam is splintered into subgroups under the general definitions Sunni and Shia, and the Islamic State views other Muslims as a more immediate enemy, because they're heretics, than other religions like Jews, Christians, Hindus, Buddhists, and Yazidis.

> Like his predecessors in [Al Qaedi in Iraq], Baghdadi favors first purifying the Islamic community by attacking Shia and other religious minorities as well as rival jihadist groups. The Islamic State's long list of enemies includes the Iraqi Shia, Hezbollah, the Yazidis (a Kurdish ethnoreligious minority located predominantly in Iraq), the wider Kurdish community in Iraq, the Kurds in Syria and rival opposition groups in Syria (including Jabhat al-Nusra).[46]

It may surprise the reader to learn that ISIS carried out several attacks against Hamas and Islamic Jihad in the Gaza Strip in 2015. ISIS publicly condemned Hamas for being too focused on the Palestinian cause, working with the Shias of Hezbollah and Iran, and not promoting a rigid enough interpretation of Islamic Law. After ISIS announced its intention to conquer Saudi Arabia in December of 2014, the Saudis considered the threat serious enough to begin building a six-hundred-mile, high-tech fence along the border with Iraq, where ISIS has a strong presence in Anbar province. Bringing the holy sites of Mecca and Medina under ISIS control would grant the caliphate more legitimacy with the world's 672 million Muslims, and some analysts believe that is, in fact, the group's objective.[47]

Now, the ability of ISIS to hold the territory it has overrun has been questioned. As of this writing, however, the only military force in the region that's consistently held its own against the Islamic State is the

Kurdish *peshmerga*. But the Kurdish resistance is hampered by opposition from the Turkish government, which wants to prevent the formation of an independent Kurdish state on its southern border.

Turkey's support of the Islamic State may be the worst-kept secret of the Syrian civil war. The Turks view their southern neighbor, Syria, as a rival, and relations between the two countries have been tense for decades. The leadership of the Kurdistan Workers Party (PKK), a group considered a terrorist organization by Turkey and the United States, operated out of the Syrian capital, Damascus, from 1978 to 1998. Syria only stopped supporting the PKK when Turkey threatened to invade. It's possible that the government of Turkey's President Recep Tayyip Erdogan justifies its support of ISIS as payback for PKK terror attacks against Turkish targets.

It's also possible that Erdogan is using the Islamic State as a cat's paw in his long game to revive the Ottoman Empire. The rise of Erdogan's Justice and Development Party (AKP) has given him the clout to roll back some of the secular reforms of Mustafa Kemal Ataturk, who modernized Turkey and aligned it with the West after the collapse of the empire in 1923. Ottoman Turkish and Arabic script will again be taught in government schools, and the number of students enrolled in state-run Islamic seminaries has grown from sixty-two thousand in 2002, when Erdogan first came to power, to over one million.[48]

Considering Erdogan's public call for Muslims to work to wrest control of Jerusalem from Israel, it's no surprise that some Middle East observers are asking in so many words: Is Turkey attempting to resurrect the Ottoman Empire on the back of the Islamic State?

Here's another uncomfortable question: Given the long tradition for timing events to coincide with major events from history for maximum symbolic impact, how far will Erdogan go to reach the ambitious economic and geopolitical goals he's set for Turkey by 2023, the one hundredth anniversary of the collapse of the Ottoman Empire?

If you've read *The Final Roman Emperor, the Islamic Antichrist, and the Vatican's Last Crusade* by Tom Horn and Cris Putnam, you know that a pseudepigraphal text called *The Apocalypse of Pseudo-Methodius*, a document falsely attributed to a fourth-century bishop named Methodius of Olympus, prophesied that a final Roman emperor would emerge in the last days who would save Christendom from the "sons of Ishmael." This false prophecy was created in the seventh century as a response to the sudden and unexpected eruption of Islam from the Arabian desert. The Christian world was reeling in shock, powerless to stop the followers of Muhammad as they overran the Holy Land, the Levant, north Africa, and most of Spain in just over a century.

But *Pseudo-Methodius* provoked an equal and opposite reaction from Islamic teachers—the Mahdi. Yes, as strange as it sounds, the savior of Islam was invented in response to the Final Roman Emperor—who was invented as a response to Islam. In other words, the Mahdi is the product of an "anything you can do, we can do better" game of oneupmanship.

Isn't it bizarre that these bits of 1,400-year-old fiction are impacting the world so powerfully today?

Another odd thing about Islamic eschatology is that neither of the major players, the Mahdi or the Dajjal, is mentioned in the Quran. All that is known about them comes from the *hadith*, collections of reports that claim to preserve the sayings of Muhammad verbatim.

Unlike the Quran, which was compiled under the authority of the early Islamic authorities in Medina, the hadith weren't collected until the eighth and ninth centuries. The sayings were not evaluated by a central authority. Islamic scholars in the centuries since have divided the hadith into *sahih* (authentic), *hasan* (strong), and *da'if* (weak).

Here's the thing: There is no universal agreement on which hadith are which. Sunni and Shia Muslims refer to different collections of hadith, and there is a small group of Quranists who reject the authority of hadith altogether.

That makes it just as difficult to compile an authoritative list of What Muslims Believe About the End Times as it is to nail down a universal set of Christian eschatological beliefs—in other words, it is impossible. Fundamental differences have divided Sunnis and Shias for more than 1,300 years and they naturally carry over into eschatology.

For example: Sunnis believe the Mahdi is yet to appear on earth while Shias believe they will see his return. These contradictory views are similar to those that distinguish the Christian Messiah from the Jewish *mashiach*.

Two other figures play key roles in Islamic eschatology. Isa, the Muslim conception of Jesus, will appear in the Last Hour to kill the Dajjal (or help the Mahdi do so), and the Sufyani, a Muslim tyrant (or national hero, depending on the sect—again, fundamental differences), will emerge in Damascus just before the Mahdi's arrival.

Isa's return, oddly enough, is prophesied in the Quran. However, Muslims believe Isa/Jesus didn't die on the cross but was taken up into heaven by Allah the same way Elijah was called up before he died. Isa, despite his miraculous birth (Muslims accept that Mary was a virgin) and rescue from the cross, will die like any other mortal man some years after his return.

The Sufyani, like the Mahdi and the Dajjal, is an apocryphal figure mentioned only in the hadith. The difference in the perception of this character in the Muslim world illustrates the depth of the hostility between Sunnis and Shias. His name stems from his ancestor, one Abu Sufyan, the leader of Muhammad's tribe who initially persecuted the self-proclaimed prophet and his followers. Although he and his family eventually converted to the faith, Abu Sufyan's son fought Muhammad's son-in-law, Ali, for control of the new Islamic empire, and eventually became the caliph.

Ali's supporters, the "Shi`at Ali" ("partisans of Ali"—later just the Shi`a), formed their own sect which persists to this day. The Shia soon began teaching that the Mahdi would return someday to defeat the champion of the Sunnis, the Sufyani, in the Levant. Many Shias today believe the Sufyani's emergence is imminent, which, of course, would be bad because he's the enemy of the Mahdi.

Conversely, some Sunnis see the Sufyani as a sort of national hero, especially in Syria, the historic homeland of the Sufyani's ancestors. Unlike Shia prophecies that portray the country in a negative light because it will be the birthplace of the Sufyani, Syria holds a place of honor for Sunnis as the site of the decisive future victory over the forces of Rome.

And, we remind you, it is also the ancestral homeland of the Amorites. That territory, the area along the Euphrates, from the ancient cities of Emar and Tuttul down to Mari, and over to the KUR MAR.TU (the mountain of the Amorites)—i.e., a wedge of Syria along the Euphrates from Lake Assad past the Islamic State's capital of Raqqa down to the Iraqi border, over to the Jebel Bishri mountain range, extending as far as the ancient oasis city of Tadmor (Palmyra, recaptured from Syrian troops by ISIS in December, 2016)—is the core of the would-be ISIS state.

This adds more fuel to an already incendiary situation. While some 670 million Muslims expect to see the Mahdi in their lifetimes, including an overwhelming majority in the Middle East and western Asia, Sunnis and Shias are looking for very different men to fulfill their prophecies.

It's a safe guess that the typical Muslim is not much better informed about his or her faith than the typical Christian. So while large majorities in nations like Syria, Iraq, and Afghanistan expect the Mahdi's return in the near future, their expectations—which are already built on teachings based on sometimes contradictory hadith of questionable authenticity—may be shaped to fit current events by charismatic and persuasive imams or political leaders.

In very broad strokes, then, we can establish some common expectations:

1. The "Last Hour" will be preceded by corruption, widespread unbelief, oppression of Muslims, declining standards of living, wars and anarchy, sexual immorality, the emergence of false prophets, and an increase in technology.

2. The armies of Rome will land at al-A'maq, a valley near Antakya (Antioch) in southern Turkey, roughly seventy miles west of Aleppo, or in Dabiq, a rural village in Syria between Aleppo and the border with Turkey. The Muslims defeat the Romans and go on to conquer Constantinople (Istanbul). This belief is at the center of the apocalyptic theology of the Islamic State. (Ever wonder

why the official news agency of ISIS is called Amaq and its official magazine is named *Dabiq*?) Not surprisingly, when Syrian rebels backed by Turkish troops and U.S. Special Forces overran Dabiq in October 2016 with virtually no resistance, the Islamic State told the faithful that the relevant *hadith* refers to a *future* battle at Dabiq, not the one they'd just lost.

3. The Dajjal emerges from the east, possibly from Khorasan, the traditional name of a region in eastern Iran and western Afghanistan, and remains on earth deceiving and oppressing people for forty days, forty months, or forty years.

4. Isa (Jesus) descends from heaven at Damascus and either helps the Mahdi kill the Dajjal or kills the Dajjal himself.

5. The Sufyani fields an army to fight the Mahdi, but the earth swallows the Sufyani and his followers before they reach the Mahdi.

6. When the fighting is over, Isa and the Mahdi will lead prayers at Jerusalem. Al-Mahdi will try to defer to Isa, but Isa will insist on remaining subordinate to the Mahdi. The two will rule over the earth for forty years before dying of old age.

Obviously, there is far more to Islamic eschatology than just the above. However, since much of it is not universally believed or deals with supernatural events that take place after the defeat of al-Dajjal, it doesn't concern us here.

And then there are multiple variations on the main theme that are unique to either Sunnis or Shias. Some Sunnis believe the Dajjal will come from Iran; some Shias believe the ongoing civil war in Syria is a sign that the end times are upon us (as do some Orthodox and ultra-Orthodox rabbis in Israel).

But Western Christians need to understand this: Both Sunnis and Shias firmly believe that the *other* sect will mistake the Dajjal for the Mahdi. News accounts of the apocalyptic beliefs of young Muslim men fighting each other in the Syrian civil war convey the sense that Sunnis and Shias are enthusiastically slaughtering each other for the privilege of going toe-to-toe with the Dajjal.

Considering that this supernatural war is over Zion, the ultimate prize, it's significant that while Mecca and Medina are the holiest sites in Islam, they play virtually no role in Muslim prophecies of the end times. Damascus and Jerusalem are far more important in Islamic eschatology. Anyone who believes this is a coincidence is, with all due respect, not paying attention. Jerusalem is the site of Yahweh's mount of assembly, and Damascus is where the apostle to the Gentiles, Paul, was brought into the faith and began his mission.

Because of the Islamic State's resilience, we must consider their interpretation of Islamic eschatology when trying to project into the future. There are important questions about some of the events on the group's prophetic timeline. For example, when ISIS refers to Rome, we must ask: Which one?

When Muhammad allegedly prophesied the future battle at al-A'maq or Dabiq, Rome wasn't Rome anymore. The Western Roman Empire ended in A.D. 476, 150 years before Muhammad's rise to power. In his day, the remnant of the Roman empire was centered on the Byzantine capital, Constantinople (modern Istanbul). And the Church in Rome didn't have anywhere near the religious influence most modern readers assume the Vatican has always possessed until at least a couple hundred years after the time of Muhammad.

So who or what is meant by Rome? Even rank and file Islamic State faithful don't agree. Some think Rome is the Roman Catholic Church, others believe it's Christendom (Western Europe), and still others the United States or Turkey—which is ironic, since evidence suggests that the governments of Turkey, the U.S., and other Western allies actively assisted the Islamic State even after it declared the caliphate.[49]

More important, as noted above, is the fundamental disagreement between Sunnis and Shias over the identity of the Mahdi. Sunnis, who comprise an estimated 87–90 percent of Muslims worldwide, have traditionally derived religious authority from the caliphate. The first caliph was appointed by the companions of Muhammad at his death because the prophet left no male heir.

Shias, however, follow the bloodline of Muhammad, believing that his true heir descends from the prophet's cousin and son by marriage, Ali. To Shias, and more specifically Twelver Shias, the Mahdi is the Twelfth Imam, Muhammad al-Mahdi, who went into hiding in A.D. 873 at the age of four.

Or so it's claimed. His father, Hasan al-Askari, lived his life under house arrest. He was apparently poisoned at the age of twenty-eight, probably by the Abbasid Caliph, and died without a male heir. That might have been the end of Shia Islam right there. But one Abu Sahl al-Nawbakhti of Baghdad saved the day by claiming that al-Askari did, in fact, have a son who had gone into *ghaybah*—"occultation" or "hiding." Like King Arthur, who will return at the hour of Britain's greatest need, the Twelfth Imam will return at the end of the age to usher in an era of peace and justice—and, of course, establish Islam as the global religion.

In short, the Mahdi's appearance will either be the arrival of a rightly guided Sunni Muslim leader, a mortal man who will rule for a time and then die, or the return of a Shiite Imam who's been supernaturally preserved for more than 1,100 years.

This is a key distinction: For Shias, the Mahdi must reappear as one specific person. In Sunni theology, "the mantle of the Mahdi can be appropriated, in the right context, by a charismatic leader megalomaniacal enough to believe Allah is directing him to wage divinely-guided jihad."[50] Here is another crucial difference: Unlike Sunnis, Shias don't believe that humans can affect the timing of the arrival of the Last Hour. The Mahdi will appear when Allah wills it and not one heartbeat sooner. In fact, they're not even supposed to fight for victorious global *jihad* until the Twelfth Imam returns.

Furthermore, there are reasons to believe that Shias, especially in Iran, aren't all that eager for the Mahdi's return. The ayatollahs, who rule about

40 percent of the world's Shia Muslims, would find the return of al-Mahdi an obstacle to maintaining their positions of authority—and with them, their comfortable lives of nice cars, pretty wives, and big houses. One Israeli scholar puts it bluntly:

> Shi`ism in general, and post-revolutionary Iranian Shi`ism in particular, is *not* only *not* messianic or apocalyptic in character, but is in fact the fiercest enemy of messianism to be found anywhere in the Muslim world or Islamic history.[51]

That may be a surprise to Americans who've heard conservative media pundits tell them for years that Twelver Shias want nothing more than to trigger the Apocalypse by destroying Israel. But since the Mahdi and Isa/Jesus are prophesied to lead the faithful in prayer at Jerusalem, the ayatollahs would seemingly prefer not to turn the Temple Mount into a radioactive crater for the Twelfth Imam.

Please understand, this is not to say that Iran is no threat to Israel and the West. Iran is a state sponsor of terror, and Americans should never forget the 241 American soldiers killed at the Marine barracks in Beirut in 1983.

Still, even if the Shias don't start it, at least one and perhaps several military battles for Zion are in our future. The Temple Mount is the prize, and its significance is recognized by Muslim, Jew, and Christian. And Sunni Muslims, whose long history of radical Mahdist movements continues with the Islamic State, generally *do* believe the Last Hour can be jump-started.

In broad terms, the eschatological beliefs of Jews and Christians are more like those of Shia Muslims than those of Sunnis, at least insofar as the influence humans can have on the timeline of the Last Days. And since Christians and Jews share the prophecies of the Hebrew prophets, it isn't surprising that there are some similarities in their eschatological teachings. However, it's in the differences, and the failure of many professing Christians to understand the basics of Bible prophecy, that danger lies.

———

Arab Muslims have controlled the Temple Mount since the conquest of Jerusalem in A.D. 638, with brief exceptions during the eleventh and twelfth centuries when European crusaders occupied the city. The Al-Aqsa Mosque, the third holiest site in Islam, and the Dome of the Rock, which sits on the spot from which Muslims believe Muhammad ascended to heaven, were constructed in the late seventh century. Now, the Grand Mufti of Jerusalem, Sheikh Muhammad Ahmad Hussein, would disagree; he's said publicly that Al-Aqsa was either built by Adam or by angels "during his time."

When Israel captured the Temple Mount during the Six-Day War in 1967, the geopolitical fallout of taking full control of Zion was considered so dangerous that the first action of Israel's Defense Minister Moshe Dayan was to take down the Israeli flag that paratroopers had raised over the mount.

The Temple Mount today is administered by the Waqf, an Islamic religious trust that has overseen the area since 1187. The government of Jordan acts as custodian of the Islamic religious sites on the mount, although security is provided by Israeli police. This arrangement, which you can guess makes no one happy, is a constant source of irritation and provocation to Jews and Muslims alike.

Christians, as spiritual descendants of Judaism, also attach special significance to the Temple Mount. In addition to the Old Testament history linked to the site, some of the major events of Jesus' life took place on the Temple Mount. As an infant, Jesus was presented at the Temple in accordance with the Law, where Simeon, a man who had been told he would live to see the Messiah, and Anna, an eighty-four-year-old prophetess, were led to Jesus by the Holy Spirit. When Jesus was twelve, He remained behind in the Temple after His parents began the journey back to Nazareth following the Passover celebration in Jerusalem. It was a full day before they discovered Jesus was missing, and at least three more before they found Him in the Temple talking with the rabbis.

Early in His ministry, Jesus visited Jerusalem during Passover, as must have been His custom, and with a zealous anger He drove the moneychangers and animal merchants out of the Temple. Later, probably during the second Passover of His ministry, Jesus healed the lame man at the Pool of Bethesda at the north end of the temple complex. Shortly before the Crucifixion, Jesus drove out the moneychangers a second time, and Matthew records that He healed many lame and blind people who came to Him at the Temple.

Still, its importance to Christians, especially in the increasingly secular West, pales in comparison to the significance of the Temple Mount to Jews. We American Gentiles can only guess at the level of frustration religious Jews must feel at being forbidden to pray on the mount.

Moshe Dayan and Israel's secular leadership in 1967 apparently believed that the mount was holy only to Muslims and nothing more than "a historical site of commemoration of the past" for Jews. By granting them access to the site, Dayan thought Jewish demands for worship and sovereignty there would be satisfied. By allowing Muslims to keep religious control of the Temple Mount, he hoped to remove the site as an inspiration for Palestinian nationalism. It was the ultimate no-win situation, as the recent wave of Palestinian violence reminds us. And it will only get worse in the years ahead.

The Jewish state is under near constant pressure from the global community to give up land for peace. Although the United States hasn't formally recognized Palestine, it's in the minority of nations that haven't done so. On November 22, 1974, the United Nations General Assembly passed a resolution recognizing the right of the Palestinian people to self-determination and independence, and the Palestine Liberation Organization was recognized as the sole legitimate representative of the Palestinian people.

The Palestinian National Council declared independence on November 15, 1988, which was acknowledged by eighty nations by the end of that year. As of late 2016, more than 70 percent of the world's 195 nations recognize Palestine as an independent state. Even the Vatican extended formal recognition in February, 2013. Israel does not, of course, maintaining that such status can only be conferred by direct negotiations between Israel and the Palestinian National Authority. To date, Israel still maintains de facto military control over the Palestinian territories.

But even in the United States, which has been Israel's strongest supporter since 1948, calls have been heard even from Republican administrations for an independent Palestine. In October of 2007, Condoleezza Rice, Secretary of State for George W. Bush, said at a news conference with Palestinian President Mahmoud Abbas:

> Frankly, it's time for the establishment of a Palestinian state. The United States sees the establishment of a Palestinian state and a two-state solution as absolutely essential for the future, not just of Palestinians and Israelis but also for the Middle East and indeed to American interests.[52]

Surprisingly, most Israelis and Palestinians say they support a two-state solution in opinion polls. However, when the compromises required to reach such a solution are spelled out, like drawing permanent borders, the status of settlements, and dividing Jerusalem, support collapses.

The Roman Catholic Church recently waded back into these troubled waters. On June 26, 2015, the Vatican signed a treaty with the "state of Palestine," essentially acknowledging the independence of a sovereign Palestine. The Israeli Foreign Ministry immediately expressed its disappointment and declared that this move would not benefit the peace process. If the Palestinian Authority can achieve independence through outside influence on Israel, why should it negotiate with the Israeli government?

News of this agreement stirred old suspicions among some Jews that the Vatican is conspiring with Palestinian leaders, and possibly with Israeli elites, to take control of the Old City and/or the Temple Mount. Stories have circulated on the Internet for years that the Vatican is working with Jewish elites on a secret deal to turn over administration of the Old City to the Roman Catholic Church.

This isn't entirely a conspiracy theory. The 1947 United Nations Partition Plan for Palestine included a proposal to designate Jerusalem *corpus separatum* (Latin for "separated body"), a zone under international control because of the city's shared religious importance. That proposal was included in the plan largely because of a powerful diplomatic effort by the Vatican, which had been concerned about the status of Christian holy sites in the Holy Land since the nineteenth century.

However, the partition plan failed. War broke out almost immediately after Israel declared its independence, and it's hard to talk when bullets are flying. Months of intense fighting left Israeli forces in control of western Jerusalem, and Israel held on to that territory when the armistice was signed that ended the 1948-49 war.

Today, at least one Middle East think tank, the Jerusalem Old City Initiative, formed by Canadian diplomats after the failure of the Camp David talks in 2000, "concluded that an effective and empowered third party presence was imperative in the Old city."[53] A similar proposal was reportedly made by the Obama administration in late 2013. US Secretary of State John Kerry, in Israel trying to broker a deal for a Palestinian state, put forward a "third party solution" for eastern Jerusalem. Under the proposal, the Vatican would control holy sites in partnership with a coalition of Muslim countries such as Turkey and Saudi Arabia. Sources close to the talks said Israeli leaders were unreceptive, especially to the idea of Turkey's participation.[54]

A suggestion that Jordan might replace Turkey in the international coalition was met with a lukewarm response in Amman. King Abdullah wasn't eager to involve his nation in a delicate and potentially explosive political situation while the Syrian civil war heated up just over his northern border.

Relations between Israel and Turkey have soured in recent years, and this is likely due to the regional ambitions of Turkey's President Recep Tayyip Erdogan. Turkey supported the so-called Gaza Freedom Flotilla, a 2010 mission to deliver construction materials and humanitarian aid to the Gaza Strip coordinated by the Free Gaza Movement and the Turkish

Foundation for Human Rights and Freedoms and Humanitarian Relief. However, since Israel and Egypt have blockaded access to the Gaza Strip since 2007, aid is normally delivered to Israel and then transferred to Palestinian authorities. The flotilla attempted to deliver the aid directly to Gaza by running the blockade.

When Israeli forces intercepted the flotilla on May 31, 2010, nine people were killed on board the Turkish ship MV Mavi Marmara. Although President Erdogan said in 2013 that relations with Israel could be normalized if certain conditions were met, in May, 2015, he called for Sunni and Shia Muslims to set aside their differences and resume efforts to assert dominance over Jerusalem, and specifically over the Temple Mount.

Turkey's deteriorating relationship with Russia after two of its F-16s shot down a Russian SU-24 fighter-bomber over Syria on November 24, 2015, gave Erdogan a new incentive to improve relations with Israel. Russia supplies more than half of Turkey's natural gas, and Vladimir Putin has shown that he's willing to use Russia's energy resources as a geopolitical weapon. Israel, meanwhile, is developing a potentially huge reserve of natural gas, the appropriately named Leviathan field, in the eastern Mediterranean.

So in June of 2016, Israel and Turkey signed a deal to normalize relations after the six-year rift. The agreement included an eventual return of ambassadors, a $20 million compensation fund for the families of those killed on the Mavi Marmara, and talks of a natural gas pipeline.

Still, at a parliamentary symposium on Jerusalem in late November 2016, President Erdogan continued his recent criticism of Israel's dealings with the Palestinians. Erdogan called on Muslims to embrace the Palestinian cause, saying that defense of the Al-Aqsa Mosque should not be left to children armed with nothing but stones.

Under current political conditions, Israel will grow increasingly resistant to pressure to relinquish control of the Temple Mount. The Netanyahu administration, which has governed Israel since 2009, has been quietly investing in efforts to prepare for the construction of the Third Temple. Those efforts include education to teach young Israelis about the importance of the Temple to the state and to Judaism, and practical efforts, mainly by a private organization called the Temple Institute, to prepare the plans, utensils, and even sacrificial animals needed to make the Temple a reality. The Israeli newspaper *Haaretz* recently disclosed that close supporters of Prime Minister Netanyahu, a deputy defense minister and a key U.S. fundraiser, had made significant financial contributions to advance the cause of the Third Temple's construction.[55]

This is laying the groundwork for a confrontation of literally biblical proportions. While modern Israel is mainly a secular society, eschatological proclamations by respected rabbis are becoming routine. Rabbi Chaim Kanievsky, considered a leading authority in mainstream Haredi (ultra-Orthodox) Judaism, not previously given to messianic predictions, has been advising Jews since 2014 to make *aliyah* (relocate to Israel) as soon as possible to prepare for Messiah's arrival. (It should be noted that Rav Kanievsky predicted the Messiah's arrival by the end of the recent Shemitah year, which concluded on September 12, 2015. Obviously that prediction was inaccurate.)

Rabbi Moshe Sternbuch, vice president of the Rabbinical Court, said in early December, 2015 that the political conflict between Turkey and Russia means that Jews should anticipate the imminent coming of the Messiah.

"We have received a direct teaching, passed down from one to another, from the Gaon of Vilna, that when Russia goes and conquers Istanbul...it is time to quickly put on your Shabbat clothes and expect the Messiah," he said.[56]

The Vilna Gaon ("Genius of Vilnius"), Elijah ben Shlomo Zalman, was an eighteenth-century rabbi and kabbalist who is considered one of the most influential rabbinical scholars since the Middle Ages. Rabbi Sternbuch is a great-grandson of the Vilna Gaon. The "direct teaching" is a prophecy that's been passed down in the family, revealed for the first time by American-born Hasidic Rabbi Lazer Brody in March of 2014:

When you hear that the Russians have captured the city of Crimea, you should know that the times of the Messiah have started, that his steps are being heard. And when you hear that the Russians have reached the city of Constantinople (Istanbul), you should put on your Shabbat (Sabbath) clothes and don't take them off, because it means that the Messiah is about to come any minute.[57]

Rabbi Brody was moved to reveal the prophecy by Russia's annexation of Crimea in March, 2014 as part of the civil war in Ukraine. Apparently, Rabbi Sternbuch and others believe this fulfilled the first part of the Vilna Gaon's prophecy.

As for the second part: As of this writing, the political situation between Turkey and Russia remains unsettled. A month after the downing of the Russian warplane, a spokesman for the Russian Ministry of Defense claimed the attack was a premeditated act by Turkey—one that may have been planned with information shared by Russia with the United States.[58] And while things have thawed between the two nations, in November of 2016, Turkey's president Recep Tayyip Erdogan declared that his troops were in northern Syria to fight ISIS and the Kurds, and to depose Syria's president Bashar al-Assad. This could bring Turkey back into conflict with the Russian troops that are in the country at the invitation of Assad.

While we give credence only to those prophecies recorded in the Bible, it's difficult to come up with an explanation for Turkey's provocative act that makes more sense than the one offered by Rabbi Sternbuch: "It is impossible to understand how a country like Turkey can start a war and refuse to apologize. They are crazy! God is confusing them, therefore we need to strengthen ourselves in repentance so we will merit a true redemption very soon."[59]

In other words, when political leaders act in irrational and self-destructive ways, supernatural forces may be guiding their decisions. That's biblical—Yahweh hardened the heart of Pharaoh in spite of the plagues that laid waste to Egypt, and He did the same to the Amorite kings Sihon and Og to bring about their doom. A day is coming with Gog of Magog will be pulled into a war by Yahweh, with supernatural hooks in his jaws. Maybe irrational behavior is what those hooks look like to the natural eye.

Further complicating matters, the Russian ambassador to Turkey, Andrey Karlov, was assassinated in Ankara December 19, 2016, by off-duty police officer Mevlut Mert Altintas. The assassin was reportedly admitted to an art gallery where Karlov was introducing an exhibit by security forces who apparently thought Altintas was a bodyguard. At this time, it's too early to say how this shocking event will affect relations between Turkey and Russia. It's worth remembering that the Turks and Russians went to war with each other a dozen times between the late sixteenth century and World War I.

Whether the Vilna Gaon's prophecy was inspired is irrelevant. What matters is that there are people in Israel who believe it, and what one believes determines what one does. And there are other religious leaders in Israel who apparently believe the end of history is fast approaching.

World events prompted Rabbi Yosef Berger, one of the rabbis in charge of King David's Tomb, to create a Torah scroll to unify all of Israel. Why? He hopes to present it to the Messiah upon His arrival. In an interview, Rabbi Berger told a reporter that he was inspired by a prophecy in chapter 3 of the book of Hosea:

> For the children of Israel shall abide many days without a king, and without a prince, and without a sacrifice, and without an image, and without an ephod, and without teraphim:
> Afterward shall the children of Israel return, and seek the LORD their God, and David their king; and shall fear the LORD and his goodness in the latter days. (Hosea 3:4–5, ESV)

Rabbi Berger said he believes that bringing Israel together with a single Torah scroll housed on Mount Zion, the site of David's Tomb and adjacent to the Temple Mount, will fulfill the prophetic goals of seeking the Lord, seeking the dynasty of David, and the construction of the Third Temple.[60]

Perhaps the most well-known recent rabbinical prediction of the Messiah is the claim by the late Rabbi Yitzhak Kaduri, a renowned kabbalist and Haredi rabbi who said he met the Messiah on November 4, 2003. Rabbi Kaduri sealed a note containing the name of the Messiah that was not to be opened until a year after his death on January 28, 2006.

The revelation that the name in Rabbi Kaduri's note was Yehoshua—Jesus—touched off just a bit of controversy.[61]

Setting aside the predictions of rabbis who see prophetic significance in the headlines, what do they expect when the Messiah finally arrives? Typically, Jews who believe in end-times prophecy think Israel will face an existential threat from a coalition of enemies invading from the north. That's the war of Gog and Magog from Ezekiel chapters 38 and 39. Christians mainly believe this war will end when God intervenes and supernaturally destroys the invading army with "torrential rains and hailstones, fire and sulfur." The Messiah plays no overt role in this battle, and Christians don't all agree over whether they'll still be on earth during this war. Some prophecy scholars believe the conflict takes place after the Rapture of the Church.

However, Jews believe the *mashiach* does participate. In fact, *two* are expected in Jewish prophecy—Mashiach ben Yosef and Mashiach ben David. Unlike the Christian understanding of the Messiah's return, Jews believe Mashiach ben Yosef and Mashiach ben David are men of this world—observant Jews rather than supernatural saviors.

The origin and character of the Messiah of the tribe of Joseph, or Ephraim, are obscure. It seems the assumed superhuman character of the Messiah was hard to reconcile with prophecies that point to His death. Jews never accepted the idea of the Messiah as a suffering servant, and obviously they wouldn't see Zechariah 12:10 ("him whom they have pierced") as Jesus. So the haggadists, writers of parables and anecdotes used to illustrate points of law in the Talmud, created a second Messiah. This figure, who would come from the tribe of Joseph or Ephraim instead of Judah, would willingly suffer for Israel and fall as a casualty in the Gog and Magog war.

In brief, Mashiach ben Yosef is killed during the Magog invasion. He's replaced and later resurrected by Mashiach ben David (or, some believe, Elijah), who goes on to purify Jerusalem, gather the Jews to Israel, build the Third Temple, reinstitute the Sanhedrin, and restore the system of sacrifices.

Then comes the final judgment: Messiah ben David judges the nations and their guardian angels, presumably the seventy placed over the nations by Yahweh after the Tower of Babel incident, and then he throws all of them—nations and angels—into Gehenna.

Of course, not all Jews believe in the literal return of the *mashiach*. Orthodox and Hasidic Jews are most likely to await his arrival, while Conservative, Reform, and Deconstructionist Jews tend to view the mashiach as nothing more than a symbol for the redemption of mankind from the evils of the world.

———◆———

As we observed earlier, surveys show that premillennial, pre-tribulational believers are in the minority among the world's Christians. There are other beliefs about end-times prophecy among people who call themselves followers of Christ that are way different.

Generally, Christians can be divided into three eschatological groups. First, premillennials believe Jesus will return to collect the faithful before He rules on earth for a thousand years. This is drawn from a literal reading of Revelation 20:1–6. Premillennials can believe in a pre-trib, pre-wrath, mid-trib, or even post-trib rapture, but regardless of the timing, Christ returns, sets up His kingdom on earth and rules for a thousand years. Then Satan is released from captivity to lead one last futile assault on Mount Zion.

The largest group of Christians is amillennial. This is the official position of most the world's organized Christian denominations. They believe the thousand-year reign of Christ is a symbolic number. Amillennials generally believe the Millennium is spiritual rather than physical, that it's underway now, and that it's identical with the current Church Age. At the end of the age, whenever that is, Christ returns to establish a permanent physical reign.

Amillennialism is the official doctrine of the Roman Catholic Church, Orthodox churches, and some Protestant denominations—mainly Lutherans, Anglicans, and Methodists. So we can't just dismiss it out of hand. But it must be asked: Looking around at the world, does it seem like Satan is bound to you?

One more thing: It's surprising, but *The Apocalypse of Pseudo-Methodius* mentioned earlier has had a strong influence on Roman Catholic eschatology. It isn't official Catholic doctrine, but some influential prophecy teachers within the Roman Catholic Church believe that the author of *Pseudo-Methodius* was on to something with the Great Monarch who arrives to save the world from Islam. The question is whether the Church or some of its faithful might try to bring him to power.

The third major prophetic category is postmillennial. Postmillennials reject a literal interpretation of Revelation 20. They put Christ's return after the Millennium, which is defined as a golden age of Christian morality. Denominations teaching a postmillennial view of prophecy generally believe the world is gradually being Christianized, and that it's our job as Christians to do it. The Millennium is explained away as an allegory or a spiritual symbol, not a literal thousand-year period.

This view grew in acceptance among Protestant churches in the West between the seventeenth and the end of the nineteenth century. During the age of industrialization, increasing prosperity (especially in the United States) and rapid technological advancement popularized the idea that the world would just get better and better until Jesus came to take over.

World War I nearly put an end to that kind of optimism. A war that kills seventeen million people in four years will do that. And then the Spanish flu epidemic of 1918–19 claimed another twenty to fifty million victims. That death toll left even solid believers disillusioned.

Over the last thirty years, though, postmillennialism has enjoyed a surge of popularity among some American charismatic Christians. Within a movement called the New Apostolic Reformation, a belief is growing that the Church must literally conquer the world for Christ. This is Dominionism, a doctrine based on the belief that mankind lost dominion in the Garden when the serpent tricked Adam and Eve, and that it's God's will for the Church to reclaim that dominion by all necessary means.

To be blunt, Dominion theology is a classic Enemy PSYOP. It twists Scripture to justify Christian jihad. First seen in charismatic circles in the second half of the twentieth century, this new teaching about the end times is so far removed from what's in the Bible that it's not even accurate to call it an interpretation of Scripture.

Emerging from the apostolic-prophetic movement, which is itself an outgrowth of the Latter Rain movement of the 1950s and '60s, the New Apostolic Reformation takes a postmillennial view of prophecy (although that's not true of everyone in this movement) and teaches a "victorious eschatology," a vision of a triumphant church that completely Christianizes the world before the return of Jesus.

Some proponents of this prophetic view teach that a select group of believers, variously called Overcomers, Joel's Army, Elijah Generation, Phinehas Priesthood, the New Breed, or Forerunners, will literally defeat the enemies of God, including the Antichrist (if he is mentioned in their teachings at all), sin, and death. In some variations, this elite group of super-Christians literally becomes the incarnate Christ.

This twisting of Revelation 12:5 takes the birth of Jesus, the "man child" born to the "woman clothed with the sun," out of the past and moves it into the future as a prophecy of His return. It also transforms Christ *into* the Church, a "corporate Christ," the Many-Membered Man-Child. This heretical doctrine, called Manifest Sons of God, is also an outgrowth of the Latter Rain movement that has somehow survived into the twenty-first century.

Furthermore, some teach that Christ, however He appears, either will not or *cannot* return until Christians physically take back dominion over the earth. This doctrine is based on a misapplication of a messianic prophecy from the Psalms:

The LORD says to my Lord: "Sit at my right hand, until I make your enemies your footstool." (Psalm 110:1, ESV)

Dominionism substitutes the Church for God or for Jesus, making Christians responsible for defeating His enemies, based on the notion that since Christ is the head of the Church, we Christians are the feet and His enemies must be subservient to us.

This is heresy. It contradicts Hebrews 2:8, Ephesians 1:20–23 and 1 Corinthians 15:24–28, which plainly show that all things are *already* "in subjection under his feet." The final enemies to be defeated are Death and Hades, and they'll be thrown into the lake of fire after the Great White Throne Judgment of Revelation 20. That is not a job for us; that responsibility belongs to Christ.

While it's undeniable that most of the world is still in bondage to the enemies of God, "He put all things under [Christ's] feet and gave him as head over all things to the church" (Ephesians 1:22). This is an "already but not yet" prophecy; Creation is already subject to Christ, and a day is soon coming when He will return to claim the earth and all that is in it.

Dominionism is a variant of replacement theology, or supersessionism, which teaches that the Church has replaced Israel in all prophecies yet unfulfilled. Perhaps to shield themselves from critics of replacement theology, a new doctrine has emerged among Kingdom Now teachers called "One New Man." It reinterprets Paul's words in Ephesians 2, describing how Christ's shed blood has destroyed the "dividing wall of hostility" between Jews and Gentiles:

> …by abolishing the law of commandments and ordinances, **that he might create in himself one new man in place of the two, so making peace**, and might reconcile us both to God in one body through the cross, thereby killing the hostility. (Ephesians 2:15–16, ESV, emphasis added)

One New Man teachers view Ephesians 2 as prophecy rather than history that was fulfilled at the crucifixion. Israel has no part in end-times prophecy, they say, not because the Church has replaced Israel, but because there will no longer *be* an Israel. A blended body of believers, neither Christian nor Jew, will take its place. This includes taking Israel's place as inheriting the land promised to Abraham's descendants in Genesis 15:18–21 and Joshua 1:3–4, the area from the River of Egypt to the Euphrates. Obviously, this would be a problem in today's world, since that area includes the Palestinian territories, Jordan, Lebanon, and parts of Syria, Iraq, Kuwait, and Saudi Arabia.

It is not a coincidence that many ministries within the apostolic-prophetic movement are focused on Israel. Drawing Jews back to the land and seeing the fulfillment of the Abrahamic land promise fits hand in glove with the belief that Christians must literally take over the world before Christ can return.

Now, to be clear, not every teacher of the One New Man doctrine wants to take over the world for Jesus. Nor does every believer in Kingdom Now understand that they're trying to replace Jesus in the end times.

But there's no way to be delicate about this: Dominionist eschatology flat-out ignores a lot of Bible prophecy, including the Rapture of the Church. It skips over the uncomfortable bits that foretell a time when the Antichrist will be "allowed to make war on the saints and to conquer them." But just like the Mahdi of Sunni Islam, modern-day prophets and apostles of the New Apostolic Reformation believe they're anointed to receive new revelations that allow them to interpret Scripture in any way they see fit.

Another case in point is the Seven Mountains Mandate, a recent teaching among American Christians that's especially relevant, considering the importance of holy mountains throughout history. This teaching grew out of a dream reportedly shared by two men, founders of prominent youth ministries, in 1977. This dream has grown over the years into a strategy for conquering the world. The Seven Mountains teaching, sometimes called the 7-M Mandate or the Seven Mountains Prophecy, holds that Christians must take control of the seven "mountains," or spheres of influence, that shape society—namely, arts and entertainment, business, education, family, government, media, and religion.

Now, there's no doubt the world would be a better place if Christians really did control those aspects of culture. And we can infer from the New Testament that Christians should try to influence the people around us on whatever "mountain" we're on. But *targeting* mountains for conquest so we can dictate morality to non-believers? That was never part of the Great Commission.

You also have to wonder why God would deliver a strategy for creating a global church by conquering seven "mountains" when prophecy specifically warns us of a future global church that's founded on seven mountains:

> And I saw the woman, drunk with the blood of the saints, the blood of the martyrs of Jesus. When I saw her, I marveled greatly.
>
> But the angel said to me, "Why do you marvel? I will tell you the mystery of the woman, and of the beast with seven heads and ten horns that carries her. The beast that you saw was, and is not, and is about to rise from the bottomless pit and go to destruction. And the dwellers on earth whose names have not been written in the book of life from the foundation of the world will marvel to see the beast, because it was and is not and is to come. This calls for a mind with wisdom: **the seven heads are seven mountains on which the woman is seated.**" (Revelation 17:6–9 ESV, emphasis added)

That may be a stretch, since we know the woman, Mystery Babylon, is a city. But in a spiritual sense, the last days church, because of its alliance with the global government of the Antichrist, will most certainly sit on those seven mountains of culture—until the ten horns ("ten kings who have not yet received royal power") and the Beast turn on it and destroy it.

While Dominion theology sounds like an outlier in modern Christianity, this relatively small group wields a surprising amount of influence with the political right in America. Elements of Dominionist doctrine have found their way into such mainstream evangelical events as the National Day of Prayer, which in 2015 called on Americans to pray for the "seven centers of influence, seven days a week." Those seven centers of influence just happened to track with the mountains of the 7-M Mandate.

Now, it's highly unlikely that Americans will ever elect an openly Dominionist president, but the influence of Dominion theology may well lead some conservative Christians to welcome a world leader just because he promises to usher in a new Judeo-Christian era.

And while this view of prophecy sounds a bit fringe, it has a lot of appeal for patriotic American Christians. Let's be honest—we're often tempted to put the Constitution before the Bible. While there is no question that the world would be a better place with more genuine Christians in government, that's not our primary calling. We are supposed to change the world from the ground up, one heart at a time, not from the top down by wielding the power of government. If that was the plan, then Jesus and the apostles wasted the greatest opportunity in history by not seizing the Roman Empire at the peak of its power.

Being good citizens means voting, serving when called, and rendering unto Caesar what is Caesar's. If we substitute politics for the Great Commission, then we've been duped by another Enemy PSYOP.

———◆———

Please remember that these are brief descriptions of the eschatological positions described above. There is a spectrum of often conflicting opinions within all of them. However, it should be enough to see that the end-times expectations of the three major monotheistic religions are widely divided, both with one another and within themselves. As events unfold, they will be interpreted differently by those religions and their imams, rabbis, ayatollahs, pastors, priests, and teachers.

As we noted earlier, the prophecies of Christ's return aren't any more specific than those of His birth. That's by God's design. Remember: The apostles who learned directly from Jesus, and the archons who opposed Him, completely misunderstood Christ's mission even after He fulfilled His mission on the cross.

And that's the point: The principalities and powers who want to destroy you and everyone you love have been hard at work for two thousand years spreading misinformation and disinformation, both inside and outside the Church.

It's the final PSYOP: Wild misinterpretations of end-times prophecy will lure Christians, Jews, Muslims, and the rest of the world into welcoming the Antichrist with open arms.

SUMMING UP

Many people, Christian, Muslim, and Jew, believe we're in the end times. But between these three monotheistic faiths, and even within them, there are wide differences of opinion over what the end times will look like.

Most Sunni Muslims, especially in the Middle East, expect the Mahdi to arrive within their lifetime. This will bring on the end of history and victorious global *jihad*. Certain groups, like the Islamic State, believe they can trigger the Apocalypse, which is basically their purpose.

Christians are divided between pre-, post-, and amillennials, distinctions between whether they believe Jesus will literally rule the earth for a thousand years before one final confrontation with Satan and the forces of evil. While most Christians don't believe there is anything we can do to change the timing of Christ's return, regardless of how they view the Millennium, a small but influential group that believes in Dominion theology is convinced that we must literally conquer the world before Jesus is allowed to leave the right hand of God.

Jews are divided on whether the *mashiach* will literally return to earth. Some of the more progressive strains of Judaism see the mashiach as a symbol of a better world. However, a growing number of Orthodox and ultra-Orthodox rabbis believe the mashiach is about to arrive—first as Mashiach ben Yosef, who will be mortally wounded in the war with Gog of Magog, and then as Mashiach ben David, who will win that war, purify Jerusalem, regather the Jews to the land, and judge the nations.

As you can see, the Enemy has a lot to work with to create the ultimate PSYOP.

10

THE GREAT INCEPTION

THE APOSTLE PAUL gave us a clear warning: Contrary to the way he's usually portrayed by Hollywood, Satan appears as an angel of light. Jesus warned us that in the last days "false christs and false prophets will arise and perform great signs and wonders, so as to lead astray, if possible, even the elect" (Matthew 24:24, Mark 13:22). So Christians have no excuse—we should be *especially* wary of anyone who claims that he, or anyone else, is the Messiah. Getting it wrong means literally siding with the devil.

Consider the following possible scenario: War erupts between Israel and its nearby neighbors. Given the long history in the Middle East, this won't be a surprise. But Daniel was told about this 2,500 years ago:

> And the king shall do as he wills. He shall exalt himself and mag-
> nify himself above every god, and shall speak astonishing things
> against the God of gods. He shall prosper till the indignation is
> accomplished; for what is decreed shall be done. He shall pay
> no attention to the gods of his fathers, or to the one beloved by
> women. He shall not pay attention to any other god, for he shall

magnify himself above all. He shall honor the god of fortresses instead of these. A god whom his fathers did not know he shall honor with gold and silver, with precious stones and costly gifts. He shall deal with the strongest fortresses with the help of a foreign god. Those who acknowledge him he shall load with honor. He shall make them rulers over many and shall divide the land for a price.

At the time of the end, the king of the south shall attack him, but the king of the north shall rush upon him like a whirlwind, with chariots and horsemen, and with many ships. And he shall come into countries and shall overflow and pass through. He shall come into the glorious land. And tens of thousands shall fall, but these shall be delivered out of his hand: Edom and Moab and the main part of the Ammonites. He shall stretch out his hand against the countries, and the land of Egypt shall not escape. He shall become ruler of the treasures of gold and of silver, and all the precious things of Egypt, and the Libyans and the Cushites shall follow in his train. But news from the east and the north shall alarm him, and he shall go out with great fury to destroy and devote many to destruction. (Daniel 11:36–44, ESV)

It is generally accepted by scholars that this section of Daniel is a prophecy of the Antichrist. Verse 36 in this chapter begins a new section of prophecies yet to be fulfilled, as verse 35 marks a break from the fulfilled prophecies of the Seleucid king, Antiochus Ephiphanes, who set up the infamous "abomination that causes desolation" in the Temple.

The account in verses 36–44 is plausible enough that it reads like a summary of one of Israel's wars from the last half-century. The main questions that need answers are the identities of the kings of the north and south, and the mysterious king of verse 36, who's identified only by the pronoun "he" for the rest of the passage.

Now, in the interest of full disclosure, many well-respected teachers of prophecy disagree with this author on that point. There isn't exactly a consensus of who "he" is, or the kings of the north and south, either. This supports our theory that God deliberately made end-times prophecy hazy so the rebellious bene elohim couldn't use it to create a deception even more convincing than the one Jesus warned us about.

This author believes the "he" in Daniel 11 from verse 36 through the end of the chapter is the Antichrist. As mentioned above, verse 36 marks a break from the first section of the chapter, which is a history of the wars between successors of Alexander the Great. In fact, Daniel's prophecy matches the events of the fourth and third centuries B.C. so well that skeptics insist the book must have been written after the fact because it reads like history. Verse 35 serves as a transition from what is history for us twenty-first-century readers to a prophecy of "the time of the end."

The king of the south may be Egypt. It could also be a coalition of nations that includes Egypt or the emergence of a powerful terrorist organization in the Sinai, where the Islamic State is just one of more than a dozen violent Islamist groups that have waged low-grade war against the Egyptian government since 2011. Although Egypt currently appears to want Israel as an ally against the Shia power in the region, Iran, it was a longtime enemy of Israel. Egypt also holds a special place in the psyche of Jews as the nation that enslaved their ancestors.

In recent history, Egypt was one of the main enemy powers in both 1967's Six-Day War and 1973's Yom Kippur War. And in 2011, when the Muslim Brotherhood ousted former president Hosni Mubarak (with the approval of the Obama administration), it became obvious that Egypt is just one regime change away from being an enemy again.

But here is another possibility: South of Israel is a wealthy kingdom that is also home to the two holiest sites of the world's largest and most deadly pagan religion. It could fulfill the prophetic roles of the king of the south and the woman in purple and scarlet of Revelation 17 and 18. We mean, of course, Saudi Arabia.

Joel Richardson builds an excellent case in his recent book *Mystery Babylon* for identifying Saudi Arabia as the harlot who rides the beast. This author is inclined to agree. The House of Saud has committed immorality with the kings of the earth and made the merchants of the world wealthy from its luxurious living. So as not to add another chapter or twelve, because building a case as thoroughly as Richardson would take another whole book, we'll just add this: Mecca, as Islam's holiest site, fits the bill of the spiritual immorality described in Revelation 17 and 18. The wealth of the kingdom, generated by a network of oil tankers that sail the oceans of the world, is known to finance terror the world over, but the wealthy elites of the West turn a blind eye as long as their coffers are filled.

The kingdom's holy cities, Mecca and Medina, are in the Hejaz, where Nabonidus, king of Babylon and servant of the moon-god Sîn, settled for ten years after his military campaign against Edom and the tribes of northern Arabia. His goal was to control the spice trade of the Near East by capturing the most strategic points at the northern end of the Incense Road. During his campaign, Nabonidus captured Tayma, Yathrib (old Medina, the early capital of Islam), and Dedan, a key oasis about 250 miles north of Medina. Dedan, probably named for the nephew of Nimrod mentioned earlier, became an independent kingdom after Babylon fell to Persia and remained free for about four hundred years until the Nabateans moved into what is now southern Jordan, around Petra, around 100 B.C.

Admittedly, that's a thin thread if we're trying to connect the moon-god to the crescent-and-star of Islam. But the lands of Dedan and Sheba in Ezekiel's day, which between them defined the main occupied areas of the Saudi kingdom today (and the part of Yemen where the Saudis have been fighting the Iran-back Houthi rebels since 2015), are mentioned in the prophecy of Ezekiel 27—the prophecy of Tyre that's linked to the lamentation over prophetic Babylon in Revelation 18.

Sheba and Dedan are also mentioned in Ezekiel 38, the prophesied war of Gog and Magog, where their merchants seem shocked at the invasion of Israel.

The Saudis have been aligned with the west since the end of World War II because that's who buys their oil. But the Saudis' covert support of violent Islamists will come to the fore at some point. Either the Islamists will throw down the House of Saud or the royal family, to save itself, will abandon the West and side with the *jihadis*. That's why, while we believe Joel Richardson is correct to identify prophetic Babylon as Saudi Arabia, we do not believe the Antichrist, the Beast on which the harlot rides, will be a Muslim—or at least he won't show himself to the world as one.

File a mental bookmark there. We'll come back.

The king of the north probably represents a Muslim coalition against Israel. Other than Egypt, invaders throughout history have traditionally attacked Israel from the north—Syria, Assyria, Babylon, Persia, Greece, and Rome all came into the Holy Land from the north. Now, remember back to our section on Ba`al's mount of assembly, Mount Zaphon—today's Jebel al-Aqra on the border between Turkey and Syria. Is it a coincidence that the phrase used to describe the location of Ba`al's palace, the uttermost parts of the north (*yerekah tsaphon*), is also the location of Magog in Ezekiel 38 and 39?

In other words, since we're looking at prophesied supernatural events (the rise of Antichrist, the Gog-Magog War, and Armageddon), prophesied characters like the king of the north might also be linked to *supernatural* north, *tsaphon*—Ba`al's mount of assembly.

In today's world, it's not hard to imagine Sunni nations to Israel's north forming an alliance as they did in 1948, 1967, and 1973—some combination of Syria, Lebanon, Iraq (or parts thereof), Saudi Arabia, Egypt, and Jordan. With the rise of Islamist sentiment in the land of Ba`al's holy mountain, Turkey must be added to that list of potential enemies. And given the apocalyptic expectations of the Sunnis in the Middle East, it's not inconceivable that these nations could unite behind a charismatic leader whose followers would declare him the Mahdi.

Naturally, Christians and Jews would be just as likely to identify this Muslim leader as the Antichrist. With all due respect to those who hold to the popular Islamic Antichrist theory, they would be wrong.

Such a coalition would seem to fulfill an apocalyptic prophecy in Psalm 83:

O God, do not keep silence;
do not hold your peace or be still, O God!
For behold, your enemies make an uproar;
those who hate you have raised their heads.
They lay crafty plans against your people;
they consult together against your treasured ones.
They say, "Come, let us wipe them out as a nation;
let the name of Israel be remembered no more!"
For they conspire with one accord;
against you they make a covenant—
the tents of Edom *[Jordan and Palestinians]* and the Ishmaelites
[Saudi Arabia],
Moab *[Jordan and the Palestinians]* and the Hagrites *[Egypt, from
Hagar, Sarah's Egyptian servant]*,
Gebal *[Hezbollah and Lebanon]* and Ammon *[Jordan and
Palestinians]* and Amalek *[Arabs in the Sinai, where ISIS now has
a presence]*,
Philistia *[Hamas and Palestinians of Gaza]* with the inhabitants of
Tyre *[Hezbollah and Lebanon]*;
Asshur *[Syria and northern Iraq]* also has joined them;
they are the strong arm of the children of Lot.
(Psalm 83:1–8, ESV, comments in brackets added)

These verses name all the Muslim nations in the immediate vicinity of Israel. If Psalm 83 does foretell a future event, Daniel 11 may describe how it unfolds.

Now, what follows will not be a popular theory among evangelical Christians, but that's exactly why it will be effective. It is the belief of this author that most analyses of eschatology feature an Antichrist whose nature is obvious even to non-students of prophecy. To be frank, the supernatural Enemy we face cannot possibly be that stupid.

In other words, if he looks like an obvious candidate to be the Antichrist, then it's not him.

Instead, a dynamic leader, the "he" of Daniel 11:36–45, will emerge at a critical hour to crush Israel's traditional enemies. He will be welcomed as a savior by Israelis and as the prophesied *mashiach* by Jews. Many Christians, especially those who expect to play some role in establishing God's millennial kingdom on earth, will see him as Christ incarnate.

This man will, in fact, be the prophesied Antichrist. And he will be an Israeli who presents himself to the world as a Jew.

This theory identifies the most logical national origin of a political figure who would build the Third Temple and reinstitute the sacrifices and offerings. (Since the Antichrist figure in Daniel 9:27 will "put an end to the sacrifice and the offering," the Third Temple, or at least a tabernacle like the tent structure used from the time of Sinai until Solomon's Temple, must be built at some point.) It's hard to imagine that an Antichrist from any other faith tradition, especially Islam, allows that to happen.

Now, the Third Temple could be built and the daily offerings resumed before the Antichrist arrives on the scene. But if he's a Muslim, what are the odds he "makes a strong covenant with many for one week" (seven years) and then waits three and a half years to stop the daily offerings on the Temple Mount?

The concept of a Jewish Antichrist—to be clear, Jewish in name only—isn't a new idea. In fact, it's a very old interpretation of end-times prophecy. Several early church fathers, including Hippolytus of Rome (A.D. 170–235) and Irenaeus (c. A.D. 130–202), believed the Antichrist would be a Jew.

> Jeremiah does not merely point out his sudden coming, but he even indicates the tribe from which he shall come, where he says, "We shall hear the voice of his swift horses from Dan; the whole earth shall be moved by the voice of the neighing of his galloping horses: he shall also come and devour the earth, and the fulness thereof, the city also, and they that dwell therein." This, too, is the reason that this tribe is not reckoned in the Apocalypse along with those which are saved. (Irenaeus, *Against Heresies,* 30.2)

Dan, of course, was one of Israel's twelve tribes. Irenaeus noted the strange omission of Dan from the list of tribes that make up the 144,000 Jews who are sealed in Revelation 7. After failing to push the Canaanites out of its allotted territory on the coast, Dan migrated north and captured the city of Laish (see Judges 18), becoming the northernmost tribe. This, by the way, is consistent with the "threat from the north" theme woven through Israel's history. Dan's territory, we remind you, extended to the southwest slope of Mount Hermon and included Banias, the Grotto of Pan.

Interestingly, Hippolytus was a disciple of Irenaeus, who was a disciple of Polycarp, who was a disciple of the apostle John. So although we can't know for sure, it's possible that the concept of a Jewish Antichrist came directly from the man who received the Revelation from Jesus Christ.

Again, this notion won't sit well with American evangelicals. We support Israel more enthusiastically than most American Jews. Please understand that the Antichrist will only *present* himself as a Jew. Given that he will seat himself in the rebuilt Temple and declare himself a god, we can be sure that he will not actually *be* a Jew. Call this the Israeli Antichrist theory, then, rather than Jewish.

You may wonder how someone like the person described by Daniel—haughty, arrogant, serving a foreign "god of fortresses"—could possibly win over Orthodox and ultra-Orthodox rabbis to be declared the *mashiach*. Good question. Here is another: How often do politicians tell people what they think instead of what they want people to hear? And how convincing a liar would someone be who is indwelt by the father of lies (John 8:44)?

You may also be wondering why the Antichrist would lead a war to destroy what would probably be his most loyal and enthusiastic followers. The answer is simple: Muslims, in the eyes of the Enemy, are already lost. Those who embrace the false teachings of Muhammad are destined for destruction. The real goal of the Fallen is to destroy the followers of Jesus Christ and the people Yahweh chose for Himself.

To put it bluntly, the best use the Enemy has for Muslims is cannon fodder. They will be a bloody sacrifice to lure Jews and Christians into worshipping the Beast, whom they will mistakenly see as a literal godsend. The destruction of a Muslim coalition would be the point in the prophetic timeline for the cosmic double-cross of Revelation 17:16–17, when the Beast and the kings of the earth betray the woman and destroy her. That would mark the transfer of the spiritual power of Islam to the Beast, and from Mecca to Jerusalem. Not by suddenly converting the world's Muslims; under the scenario we envision, the destruction of Mecca as part of a war to save Israel would be the Antichrist's means of establishing his bona fides as Messiah to the world's Christians and Jews.

Muslims in the region, where most Sunnis expect the Mahdi's arrival in the very near future, would be tempted, and probably encouraged by religious and political leaders, to view such an Israeli leader as the Dajjal. If anything, this would accelerate the trend of young Muslim men flocking to Syria to join what they think are the final battles leading to victorious global jihad.

After the decisive victory by the Antichrist, Muslim eschatology becomes irrelevant. All their teachings about the Mahdi, the Dajjal, the Sufyani, the Twelfth Imam, and so on, play no further role in human history. We don't need to worry about identifying the Mahdi because there will never *be* a Mahdi. In the view of this author, 1,400 years of Islamic theology serve only one ultimate purpose for the Enemy: Drawing Muslims into a bloody war that they will lose to establish the Antichrist's claim as the prophesied savior of Israel.

Religious Jews, who are still looking for a geopolitical mashiach, will interpret this conflict as the war of Gog and Magog, the ultimate battle in Jewish eschatology. Some Orthodox rabbis are already predicting the imminent arrival of the mashiach, and some have declared that the ongoing Syrian civil war is the war of Gog and Magog. The hero of this war would be welcomed by many Jews as their long-awaited mashiach.

And how would American Christians receive such a man? Evangelicals typically view strong Israeli leaders, such as the prime minister at the time of this writing, Benjamin Netanyahu, in a very positive light. Conservative

American Christians, especially those with a poor understanding of Bible prophecy (which is, to be honest, most of us), might support such a man's elevation to the status of a world leader.

And that is his destiny, if only for a little while.

But the Dominionist segment of Christianity, which already ignores the Antichrist in its eschatology, is looking for a Christianized version of victorious global jihad. In their world, Christ cannot return until Christians take over the world, perhaps alongside Israel as a blended body of believers called the One New Man. The apostolic-prophetic movement might just welcome this triumphant Israeli leader as Christ incarnate, a fulfillment of their false prophecy of the Manifested Sons of God. Remember, since the apostles leading this movement believe they have the authority of their biblical forebears (the men who wrote the New Testament!), they can reinterpret Scripture to make their doctrine fit current events.

Now, consider the world's reaction to the next item in the chronicle of the wars of Antichrist:

> And he shall pitch his palatial tents between the sea and the glorious holy mountain. Yet he shall come to his end, with none to help him. (Daniel 11:45, ESV)

After smashing the Muslim coalition, the Antichrist will set up his government somewhere between the Mediterranean coast and Jerusalem. But he meets an unexpected end, possibly an assassin's bullet. How could this be the Antichrist, then, if he's killed before he desecrates the Temple and declares himself a god? We look to the Revelation of John:

> And I saw a beast rising out of the sea, with ten horns and seven heads, with ten diadems on its horns and blasphemous names on its heads. And the beast that I saw was like a leopard; its feet were like a bear's, and its mouth was like a lion's mouth. And to it the dragon gave his power and his throne and great authority. **One of its heads seemed to have a mortal wound, but its mortal**

**wound was healed, and the whole earth marveled as they fol-
lowed the beast.** And they worshiped the dragon, for he had given
his authority to the beast, and they worshiped the beast, saying,
"Who is like the beast, and who can fight against it?" (Revelation
13:1–4, ESV, emphasis added)

The apparently miraculous healing of the Antichrist's head wound
will amaze the world and convince millions of his divinity.

Please notice the chilling parallel here with Jewish eschatology. Jews
expect Mashiach ben Yosef to die in the war with Magog, after which
Mashiach ben David arrives to kill the enemy leader with the breath of his
mouth and brings Mashiach ben Yosef back to life.

In some traditions, Mashiach ben David goes on to purify Jerusalem
and Israel. If the Israeli Antichrist theory is correct, that could have disturbing implications. The purification includes removing non-Jews from
Jerusalem. It's not the intention of this author to cast aspersions on our
Jewish brothers and sisters, but the Antichrist's war on the saints, prophesied in Daniel 7 and Revelation 13, *will* happen at some point, and this
seems the most likely place on the prophetic timeline.

The lessons of history aren't always pleasant, but if we take the time to
study and learn, they're usually instructive. The final break between early
Christians and their Jewish neighbors in Judea didn't occur in the first
century, as we might assume. Despite the persecution of the early church
by the Sanhedrin, Jewish Christians, which is to say *most* of the Christians
for the first hundred years or so after the Resurrection, were still part of
Jewish society. Many were still members of their local synagogues, differing from their friends and neighbors only in that they knew that Messiah
had come and was coming back. That all changed during the rebellion
against Rome led by Simon bar Kokhba.

During the Bar Kokhba revolt, which began in A.D. 132, he was
hailed as the *mashiach* by the prominent Rabbi Akiva, still considered one
of the most heroic and beloved figures in Jewish history. Bar Kokhba's
name was Simon ben Kosiba; bar Kokhba ("son of the star") was a mes-

sianic claim based on his supposed fulfillment of the prophecy of Balaam, son of Beor:

> I see him, but not now;
> I behold him, but not near:
> **a star** (*kokab*) **shall come out of Jacob,**
> and a scepter shall rise out of Israel;
> it shall crush the forehead of Moab
> and break down all the sons of Sheth.
> (Numbers 24:17, ESV, emphasis added)

Christians in Judea couldn't acknowledge a mortal man as the Messiah and refused to fight for bar Kokhba. Their punishment was confinement and death.

In the aftermath of his disastrous rebellion, he was derisively called bar Koziba—"son of the lie." Despite some stunning early success against the Romans, more than half a million Jews eventually died, leaving Judea depopulated, Jerusalem destroyed, and the land renamed Syria Palaestina—Palestine. The net result of bar Kokhba's rebellion was the end of hope for an independent Jewish state in the Holy Land for nearly two thousand years.

In recent years, however, a rise in Zionist sentiment among Israelis appears to have rehabilitated the reputation of Simon bar Kokhba. He's considered a national hero by many Israelis. Bonfires are lit on Lag Ba'Omer to celebrate his short-lived Jewish state, and Rabbi Akiva's definition of the mashiach, a geopolitical savior who looks more like Simon bar Kokhba than Jesus of Nazareth, is still the standard by which Jews evaluate claimants to the title.

This misguided nationalism can be manipulated by the supernatural Man of Lawlessness to deceive Jews into following him. He may even cite the example of bar Kokhba as he singles out Christians for destruction. This is not a criticism of Jews; we know from prophecy that the faithful will be persecuted during the period called the Great Tribulation. The Antichrist is of his father, the father of lies, and he'll use whatever lies will work. Those fooled into supporting his reverse pogrom will believe they are doing God's work—at least until "Mashiach ben David" stops the daily sacrifices and declares himself a god.

Delving into a detailed, scholarly analysis of end-times prophecy is not the purpose of this chapter or this book. Our goal is simply to show how streams of Islamic, Jewish, and Christian prophecy, especially a dangerous strain of Christian prophetic misinterpretation, may overlap and reinforce the others.

As we've already observed, we aren't going to understand the prophecies of the years to come perfectly in every detail. That's by God's design. But through study and prayer, those of us still who are still here when the prophesied Beast finally shows himself won't be victimized by a cosmic deception with eternal consequences.

SUMMING UP

Muslims are waiting for the Mahdi, but what he looks like depends on whether you ask Sunnis or Shias, and the sect they belong to within those sects. Sunni Islam, specifically radical Mahdism, may be the PSYOP the Enemy uses to lure Jews and Christians into fighting for the Antichrist.

Jews may be susceptible because they want a military/political hero, like their ancestors in the days of Jesus.

American evangelicals are vulnerable to a lie, especially a small but influential contingent of American charismatics who believe they must literally take over the world for Christ before He can return. For them, Islam is Enemy Number One in terms of numbers if nothing else. But for all of us, our natural patriotism leaves us open to the temptation to want to solve spiritual problems with geopolitics.

And the Roman Catholic Church, although it doesn't officially endorse the prophecy, looks for a day when a final Roman emperor, the Great Monarch, rises to rule over a glorious age of peace. Will the Islamic State prompt the Vatican to declare its final crusade?

Jesus said in John chapter 5, "I have come in my Father's name, and you do not receive me. If another comes in his own name, you will receive him." The Enemy's long campaign of PSYOPs, spiritual disinformation and misdirection, will lure many into making Christ's words come true.

11

RETURN TO THE MOUNTAIN

As COMPLICATED AS Bible prophecy can be, there is one aspect we can share that is sure and easy to understand. It tells us what will happen after the seals are opened, trumpets have sounded, and bowls are poured out; after the witnesses, the angels in the Euphrates, and the great battle for God's holy mountain, Armageddon.

It is, as legendary newscaster Paul Harvey would have said, the rest of the story.

The story began in Eden. Chapter 28 of Ezekiel tells us Eden wasn't only a garden, it was a mountain, and a special one. It was the seat of the gods, the home of Yahweh's divine council.

Adam and Eve were part of that council. Until that fateful day.

We all know the story, how Eve and Adam bet their future on a serpent's smooth lie. The nachash, probably one of the seraphim, the serpentine "burning ones," gambled that he'd be able to elevate himself above the throne of God.

Instead, the nachash was cast down to Sheol to become lord of the dead. Adam and Eve were kicked out of the divine council, out of the garden, and off God's holy mountain, and cherubim with a flaming sword guaranteed that their descendants would never go back there. Not on our own, anyway.

That brings us to the Paul Harvey part.

Have you noticed the language of "sonship" in the New Testament? For example:

> For you did not receive the spirit of slavery to fall back into fear, but you have received the Spirit of adoption as sons, by whom we cry, "Abba! Father!" The Spirit himself bears witness with our spirit that we are children of God, and if children, then heirs—heirs of God and fellow heirs with Christ, provided we suffer with him in order that we may also be glorified with him....
>
> For we know that the whole creation has been groaning together in the pains of childbirth until now. And not only the creation, but we ourselves, who have the firstfruits of the Spirit, groan inwardly as we wait eagerly for adoption as sons, the redemption of our bodies. (Romans 8:15–17, 22–23, ESV)

That's a remarkable claim by Paul. We are adopted as sons (and by "sons," he means sons and daughters), which makes us "fellow heirs with Christ," entitled to use the intimate term *abba* when we call on our heavenly Father. Daddy!

Now, being adopted into the royal family as joint heirs with Christ doesn't guarantee an easy life. As Jesus suffered, so might we. Jesus warned us that we'd have tribulation in this world (John 16:33). That means trouble, whether we're here for the Great Tribulation or not. Anyone who tells you that following Jesus guarantees a happy, healthy, wealthy life is selling something.

But what God has ahead for us is so much better than anything we'll endure in this life. As Paul wrote, "this light momentary affliction is preparing for us an eternal weight of glory beyond all comparison" (2 Corinthians 4:7).

In his letter to the Galatians, Paul takes the sonship language a step further. He compares our relationship to God through Christ to our former relationship with Him through the Mosaic Law. It's as different as the relationships between a father and a slave and a father and his son.

> Now before faith came, we were held captive under the law, imprisoned until the coming faith would be revealed. So then, the law was our guardian until Christ came, in order that we might be justified by faith. But now that faith has come, we are no longer under a guardian, for in Christ Jesus you are all sons of God, through faith…. And if you are Christ's, then you are Abraham's offspring, heirs according to promise.
>
> I mean that the heir, as long as he is a child, is no different from a slave, though he is the owner of everything, but he is under guardians and managers until the date set by his father. In the same way we also, when we were children, were enslaved to the elementary principles of the world. But when the fullness of time had come, God sent forth his Son, born of woman, born under the law, to redeem those who were under the law, so that we might receive adoption as sons. And because you are sons, God has sent the Spirit of his Son into our hearts, crying, "Abba! Father!" **So you are no longer a slave, but a son, and if a son, then an heir through God.** (Galatians 3:23–26, 29; 4:1–7, ESV, emphasis added)

The Law given to Moses was our guardian and manager until the time set by God. Christ has redeemed us and brought us out of that captivity so we can be adopted into the family—or, more accurately, brought back into the family. Because that's been God's plan, and Christ's mission, from the beginning.

Do yourself a favor: Read the first two chapters of Hebrews. They are truly remarkable—awe-inspiring, if you stop and think about them. I will excerpt the main verses below with minimal comment, because there is absolutely nothing I can add to their beauty and power.

Long ago, at many times and in many ways, God spoke to our
fathers by the prophets, but in these last days he has spoken to
us by his Son, whom he appointed the heir of all things, through
whom also he created the world. He is the radiance of the glory
of God and the exact imprint of his nature, and he upholds the
universe by the word of his power. After making purification for
sins, he sat down at the right hand of the Majesty on high, having
become as much superior to angels as the name he has inherited is
more excellent than theirs.

For to which of the angels did God ever say,
"You are my Son,
today I have begotten you"?
Or again,
"I will be to him a father,
and he shall be to me a son"?
And again, when he brings the firstborn into the world, he
says,
"Let all God's angels worship him."
Of the angels he says,
"He makes his angels winds,
and his ministers a flame of fire."
But of the Son he says,
"Your throne, O God, is forever and ever,
the scepter of uprightness is the scepter of your kingdom.
You have loved righteousness and hated wickedness;
therefore God, your God, has anointed you
with the oil of gladness beyond your companions."
And,
"You, Lord, laid the foundation of the earth in the beginning,
and the heavens are the work of your hands;
they will perish, but you remain;
they will all wear out like a garment,
like a robe you will roll them up,

like a garment they will be changed.
But you are the same,
and your years will have no end."
And to which of the angels has he ever said,
"Sit at my right hand
until I make your enemies a footstool for your feet"?
Are they not all ministering spirits sent out to serve for the sake of those who are to inherit salvation?...

For it was not to angels that God subjected the world to come, of which we are speaking. It has been testified somewhere,
"What is man, that you are mindful of him,
or the son of man, that you care for him?
You made him for a little while lower than the angels;
you have crowned him with glory and honor,
putting everything in subjection under his feet."

Now in putting everything in subjection to him, he left nothing outside his control. At present, we do not yet see everything in subjection to him. But we see him who for a little while was made lower than the angels, namely Jesus, crowned with glory and honor because of the suffering of death, so that by the grace of God he might taste death for everyone.

For it was fitting that he, for whom and by whom all things exist, in bringing many sons to glory, should make the founder of their salvation perfect through suffering. For he who sanctifies and those who are sanctified all have one source. That is why he is not ashamed to call them brothers, saying,
"I will tell of your name to my brothers;
in the midst of the congregation I will sing your praise."
And again,
"I will put my trust in him."
And again,
"Behold, I and the children God has given me."
(Hebrews 1:1–14, 2:5–13, ESV, emphasis added)

The author of Hebrews establishes that Jesus is far superior to any angel, even his brothers, the faithful sons of God who continued to serve as Yahweh's divine council through the long years of humanity's banishment from God's holy mountain.

But a day is coming. Oh, what a day! It was made possible because the founder of our salvation, Christ, "for a little while was made lower than the angels" so that he could save us from death.

He walked among us. He knows our trials and our weaknesses. He suffered like us, and He was subject to the same temptations, disappointments, and betrayals as us.

Did I say the same? No, that's not right. Who can imagine the depth of His heartache when Jesus turned to look at Peter after the disciple denied Him a third time?

And yet we are told that because He who sanctifies, Jesus, has the same source—God—as we who are sanctified, He is not ashamed to call us brothers and sisters.

On that glorious day when Jesus returns to bring us home, He will announce to the gathered assembly, "Behold, I and the children God has given me."

And then, in the midst of the divine council on Yahweh's mount of assembly, *Jesus Christ will sing our praise!*

Can you even *imagine* it? THAT will be a family reunion for the ages!

12

THE LONG WAR

LET'S SUM UP.

A long war is underway. It's been waged by the gods against Yahweh since Eden, and maybe even before. Most of the battles take place beyond the limits of our perception. Sometimes, though, the war spills over into the natural realm in a spectacular way.

However, the Bible has been desupernaturalized by well-meaning pastors and teachers. Events of spiritual significance have been stripped of their cultural and religious context to the point that even Christian scholars look for natural phenomena to explain battles in the spirit realm.

Events that have no natural explanation, like the Transfiguration or the Red Sea crossing, are framed as myths or examples of extreme faith. The worldview of the men who wrote the books of the Bible is missing, and so we don't process the events as spiritual warfare, battles between supernatural entities—God vs. the gods.

This war is a rebellion by supernatural beings who want to establish a mount of assembly above God's. That is why mountains have been the site of so many key events in this war. To recap:

1. **Eden**. The garden of Yahweh, the seat of the gods, and the holy
   mountain of God. It was here where the nachash, possibly one
   of the seraphim, lured the first humans into disobedience. This
   allowed death into the world and cost Adam and Eve, and all of
   humanity to the present day, a place in the divine council.

2. **Hermon**. The mountain on the border between modern-day
   Syria and Israel where two hundred Watchers made a pact to corrupt humanity. They not only took wives, producing the monstrous Nephilim, but they also taught mankind forbidden knowledge. For this sin, the Watchers have been chained in darkness
   until the prophesied judgment. To cleanse the earth of the human
   and animal bloodlines corrupted by the Watchers, God sent the
   Flood. Noah and his family were saved to repopulate the earth.
   The biblical Flood story shares too many parallels with earlier
   Mesopotamian accounts to be coincidence. The myths of Sumer
   and Babylon include stories of divine sages called *apkallu* who
   were undoubtedly the Watchers. Babylonian myth also remem-
   bers Mount Hermo as the "abode of the gods." while Amorite
   cultures in the Levant believed Hermon was El's mount of assembly, where he held court with Ba`al (the storm-god Hadad) and
   the seventy sons of El.

3. **Babel**. This was mankind's attempt to force its way back into the
   divine council by building an artificial mount of assembly. Based
   on the historic evidence, we believe that Nimrod was the Sumerian king Enmerkar, who ruled an empire from the city of Uruk.
   The tower was the temple of Enki, lord of the earth, at the ancient
   city of Eridu. Enki's home was the *abzu* below the temple, a fresh-
   water aquifer necessary for life. *Abzu* is the word from which
   we get the word "abyss." Enmerkar/Nimrod is remembered in
   Sumerian myth as the king who tried to rebuild and expand the
   E-abzu (House of the Abyss) to create "the abode of the gods."
   during which the single language of men was confused into many
   tongues.

4. **Sinai**. Yahweh called Moses from Egypt to the Mountain of God and charged him with bringing His people out of Egypt. The ten plagues of Egypt included the slaying of the firstborn, a night on which God executed judgment on all the gods of Egypt. The crowning event was the crossing of the Red Sea, which was deliberately orchestrated to demonstrate Yahweh's power over the chief god of the rulers of northern Egypt at that time, Ba`al. Afterward, God invited Moses, Aaron, and the seventy elders of Israel to climb Mount Sinai to eat a meal with Him, the first time humans had been face to face with God on His holy mountain since Eden. This also sent a message to the Fallen who had passed themselves off as the seventy sons of El: Yahweh had chosen His people and a day was coming when they would be restored to the divine council.

5. **Zaphon**. This mountain, Jebel al-Aqra on the Mediterranean coast near the border between Turkey and Syria, was the site of Ba`al's palace. It was also sacred to Zeus, Jupiter, and the Hittite and Hurrian storm-gods, Tarhunt and Teššub, all of them aspects of the Amorite storm-god, Hadad/Addu, known in the Bible as Ba`al. The Hebrew prophets knew very well that Ba`al called Zaphon his home. Several references by the prophets and in the Psalms make direct comparisons between Zaphon and Yahweh's holy mountain, Zion. It was so important in Israel that the name of Ba`al's mountain, *tsaphon*, became the Hebrew word for the compass point north.

6. **Carmel**. More than five hundred years after the conquest of Canaan, Ba`al had encroached into Israel. The northern tribes had rebelled against the House of David and Israel's first king established the pagan worship of golden calves at Bethel and Dan—probably the worship of Ba`al, who was often represented by a bull or calf. By the time of Elijah, Ahab, and Jezebel, Mount Carmel marked the boundary between Israel and Phoenicia, the descendants of the Amorite-Canaanite kingdoms of the second

millennium B.C. The confrontation on Carmel, like the crossing
of the Red Sea, was specifically orchestrated by Yahweh to demonstrate His power over things that were supposed to belong to
Ba`al. Furthermore, the miracle on Mount Carmel reminded the
Fallen that Israel was Yahweh's allotted heritage.

7. **Zion**. The war over Israel and Mount Zion was personal to Jesus.
He performed many miracles on the Temple Mount, twice chased
the moneychangers out of the temple, and wept over the fate of
Jerusalem. Much of His ministry was devoted to healing the sick
and casting demons out of people, a cleansing of the land that was
again a message directed at the Fallen: *Get off my land!* Jesus sent
an even more direct message to the Fallen: The Transfiguration
on the slopes of Mount Hermon, El's mount of assembly, Peter's
confession of faith in front of the Grotto of Pan, and the sending
of the seventy (or seventy-two) disciples into Galilee immediately
afterward could have been written in blazing letters across the sky:
Here I am. I am reclaiming what is mine. Your days are numbered.

8. **Jabal al-Nour**. The Fallen didn't just roll over. At a mountain
outside Mecca, an entity that claimed to be the angel Gabriel
delivered a series of messages to Muhammad, a man of obvious
charisma and ability. Islam emerged from the Arabian desert to
become what may be the largest religion on earth within the next
half century. Compelling evidence links Islam to older pagan gods
of the ancient Near East, especially the moon-god, who has been
worshiped since at least the third millennium B.C. It may not
be a coincidence that the first city Yahweh directed the Israelites
to attack in Canaan was Jericho—a cult center devoted to, and
named for, the moon-god.

The Bible tells us the world will someday be consumed by a conflict that
will be the war to end all wars. We can't know exactly how it plays out, and

that's at least partly by God's design. Just as He obscured the prophecies of Jesus' first coming to confuse the Enemy, He's concealed enough about Christ's return to prevent the Fallen from fabricating a convincing lie.

There is one thing of which we can be sure. If anyone claims to be Christ, look at his feet. If they're touching the ground, he's lying. (Prophecy tells us Jesus returns in the clouds with great glory. If we're here when that happens, it will be impossible to miss.)

We also need to remember that the Enemy wants us feuding and fighting over anything and everything. When we're fighting, we're not making disciples. It's hard to convince people that God is love when they see us tearing each other apart or coming at them with hatred in our eyes. Let's not make ourselves stumbling blocks between unbelievers and the gospel. Our public behavior won't serve as a valid excuse when they try to justify their disbelief to the ultimate judge, but hell is hell whether you have a good reason to be there or not.

Bible prophecy can make mild-mannered believers angry enough to question the faith of our brothers and sisters in Christ. Since we're not going to understand the prophecies of Christ's return better than the apostles grasped the ones of His first coming, we should probably take a few deep breaths and get over ourselves. Let God explain it to us on the way up.

Meanwhile, let's dig deeper into God's word to guard ourselves against the PSYOPs of the Fallen. We're behind enemy lines, and every day we're surrounded by propaganda. And sadly, at least in part because we've desupernaturalized the history of the war, the Bible, we often forget that our fight isn't against human opponents.

Put on the whole armor of God, that you may be able to **stand** against the schemes of the devil. For we do not wrestle against flesh and blood, but against the rulers, against the authorities, against the cosmic powers over this present darkness, against the spiritual forces of evil in the heavenly places. Therefore take up the whole armor of God, that you may be able to withstand in the evil

day, and having done all, to **stand** firm. **Stand** therefore, having
fastened on the belt of truth, and having put on the breastplate
of righteousness, and, as shoes for your feet, having put on the
readiness given by the gospel of peace. In all circumstances take up
the shield of faith, with which you can extinguish all the flaming
darts of the evil one; and take the helmet of salvation, and the
sword of the Spirit, which is the word of God. (Ephesians 6:11–
17, ESV, emphasis added)

These verses, written by Paul to the church at Ephesus, define our
role on the cosmic battlefield. The armor he described would have been
familiar to his readers; everyone in the first century A.D. knew what a
Roman soldier looked like.

But the key word in that passage is the one I've put in bold type:
Stand. *Stand.* Yes, a war is coming. Our mission is not to conquer the
earth, but to hold our ground—to stand, and stand firm, until the King
returns.

Paul was a learned man. He knew the difference between "stand" and
"advance." He chose his words with care—and remember, he was under
the influence of the Holy Spirit. We are not conquerors, not until Christ
returns at the head of a heavenly army. For now, we stand—and work
with all our might, while the battle rages around us, to protect the weak
and rescue the wounded.

This is a literal war against intelligent evil. The old gods want to kill
us and everyone we love. We are surrounded. We're outnumbered. But the
cavalry is coming with our King at the head.

With the permission of my loving wife, Sharon, I will here excerpt a
prayer she recently wrote for Baxter, a wonderful character in her *Redwing
Saga* series of novels. This prayer is in a frame on the wall of my office.
Though these are words in a work of fiction, I have never read or heard
anything as inspired that isn't between the covers of a Bible.

May this be how we see, and pray for, our fellow soldiers in Christ:

The giant servant with the servant's heart bowed his head. "Then, let us pray," he said, his sonorous voice low.

"Our Gracious and wonderful Saviour and King, we humbly come to your throne this morning, many of us weighed down with the troubles of the day already, some with health issues, some with financial woes, others with concerns about family and friends, some carrying fear and doubt that darkens the soul and devours the mind—but no matter what our worries, no matter what our concerns or our anxieties, no matter the dross or decay of the world that rises to the top and surrounds, they all vanish and are vanquished in the light of Your face!

"It is unusual for me to speak before such an honourable gathering of fellow soldiers. I say this not because most of these men bear noble titles, but because they bear noble scars. Physical as well as spiritual. I have watched the men of this circle rush into battle wearing naught but your promises upon their mortal frames, wielding nary an ax, but flashing the sword of the spirit in the eyes of the enemy. A double-edged blade likened unto the Word of God that proceeds out of them with a fiery vengeance! But as brave as these deeds are—as valiant as their exploits in armour might be, this circle's greatest achievements are accomplished not on the battlefield but within the quietness of the prayer closet. Before these men take their feet, they spend time on their knees. Every one of them, and it is my honour to be called their fellow.

"I know not what plans the enemy now devises, my Lord, but You do. Help us to unmask their faces and uncover their secrets. Let us rise to the fight until our arms fail and our breath be gone. Let us crawl when our legs will no longer run, whisper when our voices fail, feel our way forward when our eyes become blind. Bind upon our hearts a love for one another that knows no language other than love and knows no title other than brother or sister. Help us to serve you with all our strength, soul, and mind until the end of our days,

and when those days are done, to continue encouraging our beloved fellows whilst we kneel before your throne.

"As St. Paul wrote, 'Wherefore seeing we also are compassed about with so great a cloud of witnesses, let us lay aside every weight, and the sin which doth so easily beset us, and let us run with patience the race that is set before us, Looking unto Jesus the author and finisher of our faith; who for the joy that was set before him endured the cross, despising the shame, and is set down at the right hand of the throne of God.'

"Saviour and King, I shall consider it a privilege to be counted amongst that cloud of witnesses one day, standing alongside men like Paul and Peter and James, cheering on this group of men who continue to battle upon the field. May that day come for us all when our Saviour returns. In the meantime, we consider it all joy to partake of only a small share of that which you endured on our behalf. Thank you for each man here. In the name of our King, I ask it. Even Christ Jesus. Amen."

———

We don't know exactly how events will unfold in the years ahead. Jesus warned us of deception that would fool even the elect, if it were possible. Many will be misled and fall victim to Satan's final PSYOP before his inevitable end.

Pray for wisdom and discernment. Love God, and love your neighbor as yourself. And look forward to that day when Christ Himself leads us back to God's holy mountain—and sings our praise in the divine council to welcome us home.

BIBLIOGRAPHY

Adherents.com. n.d. *Major Religions of the World Ranked by Number of Adherents.* Accessed February 28, 2016. http://www.adherents.com/Religions_By_Adherents.html.

Adler, Dr. Rivkah Lambert. 2015. *18th Century Jewish Mystics Predicted Future Conflict Between Russia, Turkey Ripe with Messianic Implications.* November 29. Accessed December 27, 2016. http://www.breakingisraelnews.com/54943/200-years-ago-war-between-turkey-russia-prophesied-sign-redemption-jewish-world/.

—. 2015. *Rabbi Kaduri "Jesus as Messiah" Claim Proven as False.* June 17. Accessed December 28, 2016. https://www.breakingisraelnews.com/43554/rabbi-kaduri-jesus-as-messiah-claim-discredited-as-false-jewish-world/#OosqSmOpzVQcbYfS.99.

Allon, Niv. 2007. "Seth is Baal—Evidence from the Egyptian Script." In *Egypt and the Levant,* by Manfred Bietak, 15–21. Wien: Österreichische Akademie der Wissenschaften.

Annus, Amar. 1999. "Are There Greek Rephaim? On the Etymology of Greek Meropes and Titanes." *Ugarit-Forschungen* 31: 13–30.

—. 2010. "On the Origin of Watchers: A Comparative Study of the Antediluvian Wisdom in Mesopotamian and Jewish Traditions." *Journal for the Study of Pseudepigrapha* 19 (4): 277–320.

—. n.d. *The Suteans and the Sons of Seth (1).* Accessed December 20, 2016. http://www.aakkl.helsinki.fi/melammu/database/gen_html/a0001500.php.

Barna Group Inc. 2009. *Barna Survey Examines Changes in Worldview Among Christians over the Past 13 Years.* March 9. Accessed December 22, 2016. https://barna.org/barna-update/transformation/252-barna-survey-examines-changes-in-worldview-among-christians-over-the-past-13-years.

Berkowitz, Adam Eliyahu. 2015. *Prominent Rabbis Sternbuch, Amar Hint That the Messiah is "Just Around the Corner."* December 9. Accessed December 28, 2016. http://www.breakingisraelnews.com/55777/turkeysyria-conflict-unfolding-prominent-rabbis-hint-messiah-around-corner-jewish-world/.

—. 2015. *Torah Scroll Being Written to Present to Messiah Upon His Arrival.* December 15. Accessed December 28, 2016. http://www.breakingisraelnews.com/56244/rabbi-sets-out-mission-write-torah-scroll-present-%20messiah-upon-arrival-jewish-world/.

Bietak, Manfred. 2010. "From Where Came the Hyksos and Where Did They Go?" In *The Second Intermediate Period (Thirteenth-Seventeenth Dynasties): Current Research, Future Prospects*, edited by Marcel Marée, 139–181. Leuven: Peeters.

—. 2005. "The King and the Syrian Weather God on Egyptian Seals of the Thirteenth Dynasty." *CASAE* 34 (1): 201–212.

Black, J. A., Cunningham, G., Fluckiger-Hawker, E, Robson, E., and Zólyomi, G. 2003. *A cir-namcub to Inana (Inana I).* June 1. Accessed December 17, 2016. http://etcsl.orinst.ox.ac.uk/cgi-bin/etcsl.cgi?text=t.4.07.9&charenc=j#.

—. 2003. *Enmerkar and the Lord of Aratta.* June 6. Accessed

December 17, 2016. http://etcsl.orinst.ox.ac.uk/cgi-bin/etcsl. cgi?text=t.1.8.2.3#.

—. 2001. *The Sumerian King List: Translation.* July 9. Accessed December 24, 2016. http://etcsl.orinst.ox.ac.uk/section2/tr211.htm.

Blau, Uri. 2015. *Netanyahu Allies Donated to Groups Pushing for Third Temple.* December 9. Accessed December 28, 2016. http://www. haaretz.com/settlementdollars/1.690821.

Bodi, Daniel. 2014. "Is There a Connection Between the Amorites and the Arameans?" *Aram* 26 (1 & 2): 383–409.

Boer, Rients de. 2014. *Amorites in the Early Old Babylonian Period.* PhD Thesis, Department of Humanities, Leiden University, Leiden: Leiden University.

Burton, Dr. Judd H. 2010. *Religion, Society, and Sacred Space at Banias: A Religious History of Banias/Caesarea Philippi, 21 BC–AD 1635.* PhD Thesis, Department of Philosophy, Texas Tech University, Lubbock: Texas Tech University.

Byman, Daniel L. and Williams, Jennifer R. 2015. *ISIS vs. Al Qaeda: Jihadism's Global Civil War.* February 24. Accessed December 28, 2016. https://www.brookings.edu/articles/ isis-vs-al-qaeda-jihadisms-global-civil-war/.

Center for International Policy Studies. 2015. *Jerusalem Old City Initiative.* October 20. Accessed December 28, 2016. http://www. cips-cepi.ca/event/jerusalem-old-city-initiative/.

Chiera, Edward, and Samuel N. Kramer. 1934. *Sumerian Epics and Myths.* Chicago, IL: The University of Chicago Press.

Coon, C.S. 1949. "The Eridu Crania: A Preliminary Report." *Sumer 5* 103-06.

Cross, Frank Moore. n.d. *Canaanite Myth and Hebrew Epic: Essays in the History of the Religion of Israel.* Cambridge, MA: Harvard University Press.

Daems, Aurelie. 2010. "A Snake in the Grass: Reassessing the Ever-Intriguing Ophidian Figurines." In *Beyond the Ubaid: transformation and integration in the late prehistoric societies of the Middle East,* edited by Robert A. Carter and Graham Philip, 149-161. Chicago, IL: Oriental Institute of the University of Chicago.

Furnish, Dr. Timothy R. 2011. "A Western View on Iran's WMD Goal: Nuclearizing the Eschaton, or Pre-Stocking the Mahdi's Arsenal?" *Institute for Near East & Gulf Military Analysis.* January. Accessed December 28, 2016. http://www.inegma.com/Admin/Content/File-8102013131623.pdf.

—. 2012. *Mahdism (and Sectarianism and Superstition) Rises in the Islamic World.* August 13. Accessed December 28, 2016. http://historynewsnetwork.org/article/147714.

—. 2016. *Sects, Lies, and the Caliphate: Ten Years of Observations on Islam.* Timothy R. Furnish, PhD.

—. 2015. *Ten Years' Captivation with the Mahdi's Camps: Essays on Muslim Eschatology, 2005–2015.* Timothy R. Furnish, PhD.

Güterbock, Hans Gustav. 1948. "The Hittite Version of the Hurrian Kumarbi Myths: Oriental Forerunners of Hesiod." *American Journal of Archaeology* 1952 (41): 123–134.

Gadd, C .J. 1958. "The Harran Inscriptions of Nabonidus." *Extrait d'Anatolian Studies* VIII: 35–92.

Gardiner, Alan H. 1932. *Late Egyptian Stories.* Vol. 1. Bruxelles: Bibliotheca Aegyptiaca.

Goldwasser, Orly. 2006. *King Apophis of Avaris and the Emergence of Monotheism.* Vol. 2, in *Timelines,* edited by E. et al. Czerny, 129–133. Leuven: Peeters.

Goodgame, Peter D. 2012. *The Second Coming of the Antichrist.* Crane, MO: Defender Publishing.

Gordon, Cyrus H. 1958. "Abraham and the Merchants of Ura." *Journal of Near Eastern Studies* 17 (1): 28–31.

Harris, Rivkah. 1991. "Inanna-Ishtar as Paradox and a Coincidence of Opposites." *History of Religions* 30 (3): 261–278.

Hays, Christopher B. 2008. "*I Set Before You Death and Life*": The

Rhetoric of Death in First Isaiah. PhD Thesis, Division of Philosophy, Emory University, Atlanta: Emory University.

Heiser, Dr. Michael S. n.d. *The Nephilim.* Accessed December 16, 2016. http://www.sitchiniswrong.com/nephilim/nephilim.htm.

—. 2015. *The Unseen Realm: Recovering the Supernatural Worldview of the Bible.* Bellingham, WA: Lexham Press.

Horn, Thomas. 2013. *Zenith 2016: Did Something Begin in the Year 2012 That Will Reach Its Apex In 2016?* Crane, MO: Defender Publishing.

Horn, Thomas, and Cris Putnam. 2016. *The Final Roman Emperor, the Islamic Antichrist, and the Vatican's Last Crusade.* Crane, MO: Defender Publishing.

Klein, Aaron. 2013. *International mandate to control sections of Israel's capital.* December 15. Accessed December 28, 2016. http://www.wnd.com/2013/12/u-s-plan-gives-jerusalem-holy-sites-to-vatican/.

Kramer, Samuel N. 1968. "The Babel of Tongues: A Sumerian Version." *Journal of the American Oriental Society* 88.

Lönnqvist, Minna. 2006. "Archaeological Surveys of Jebel Bishri: The Preliminary Report of the Finnish Mission to Syria, 2000–2004." *Kaskal* 3: 203–240.

Lake, Dr. Michael. 2016. *The Sheeriyth Imperative: Empowering the Remnant to Overcome the Gates of Hell.* Crane, MO: Defender Publishing.

Lee, Matthew. 2007. *Rice: Now is Time for Palestinian State.* October 15. Accessed December 28, 2016. http://www.washingtonpost.com/wp-dyn/content/article/2007/10/15/AR2007101500703_pf.html.

Lewis, Theodore J. 2014. "Feasts for the Dead and Ancestor Veneration in Levantine Traditions." In *In Remembrance of Me: Feasting with the Dead in the Ancient Middle East,* edited by Virginia Rimmer Herrmann and J. David Schloen, 69–74. Chicago: The Oriental Institute.

Lindquist, Maria. 2011. "King Og's Iron Bed." *The Catholic Bible Quarterly* 73: 477–492.

Lipinski, Edward. 1971. "El's Abode: Mythological Traditions Related to Mount Hermon and to the Mountains of Armenia." *Orientalia Lovaniensa Periodica* 2: 13–69.

Livingston, Dr. David. n.d. *The Fall of the Moon City.* Accessed December 17, 2016. http://davelivingston.com/mooncity.htm.

Lorentz, Kirsi O. 2010. "Ubaid Headshaping: Negotiations of Identity Through Physical Appearance?" In *Beyond the Ubaid: Transformation and Integration in the Late Prehistoric Societies of the Middle East*, edited by Robert A. Carter and Graham Philip, 125-148. Chicago, IL: The Oriental Institute of the University of Chicago.

Münnich, Maciej M. 2009. "Two Faces of Resheph in Egyptian Sources of the New Kingdom." In *Iconography and Biblical Studies; Proceedings of the Iconography Sessions at the Joint EABS/SBL Conference, 22-26 July 2007, Vienna, Austria*, edited by Izaak J. de Hulster and Rüdiger Schmitt, 53–66. Münster: Ugarit-Verlag.

Maghen, Ze'ev. 2008. "Occultation in Perpetuum: Shi`ite Messianism and the Policies of the Islamic Republic." *Middle East Journal* 62 (2): 232–257.

Mello, Alex and Knights, Michael. 2016. *Losing Mosul, Regenerating Diyala: How the Islamic State Could Exploit Iraq's Sectarian Tinderbox.* October 25. Accessed December 22, 2016. https://www.ctc.usma.edu/posts/losing-mosul-regenerating-in-diyala-how-the-islamic-state-could-exploit-iraqs-sectarian-tinderbox.

Missler, Dr. Chuck. 2001. *The Military Book of the Old Testament: The Conquering Hero—Joshua.* March 19. Accessed December 17, 2016. http://www.khouse.org/articles/2001/319/.

Ortega, Christopher E. 2015. "Inanna: Reinforcer of Heteronormativity, or Legitimizer of Non-Heteronormativity?" Conference presentation, Department of Religion, California State University, Long Beach.

Ostroukh, Andrey. 2015. *Russia Suggests Turkey Planned Downing of Warplane.* December 23. Accessed

December 28, 2016. http://www.wsj.com/articles/
russia-suggests-turkey-planned-downing-of-warplane-1450899609.

Pew Research Center. 2011. *Global Christianity—A Report on the Size and Distribution of the World's Christian Population.* December 19. Accessed December 22, 2016. http://www.pewforum. org/2011/12/19/global-christianity-exec/.

Pritchard, James B., ed. 1958. *The Ancient Near East: Volume I, An Anthology of Texts and Pictures.* Princeton, NJ: Princeton University Press.

Rendsburg, Dr. Gary. n.d. *YHWH's War Against the Egyptian Sun-God Ra: Reading the Plagues of Locust, Darkness, and Firstborn in their Ancient Egyptian Context.* Accessed December 24, 2016. http:// thetorah.com/yhwhs-war-against-the-egyptian-sun-god-ra/.

Richardson, Joel. 2017. *Mystery Babylon.* Los Angeles, CA: WND Books.

Rohl, David M. 1999. *Legend: The genesis of civilisation.* London: Arrow.

—. 2001. *The Lost Testament: From Eden to Exile: The Five-Thousand Year History of the People of the Bible.* London: Century.

Safar, Fu'ād, Seton Lloyd, and Muḥammad 'Alī Muṣṭafá. 1981. Eridu. Baghdad: Republic of Iraq, Ministry of Culture and Information, State Organization of Antiquites and Heritage.

Schmidt, Brian B. n.d. *Israel's Beneficent Dead: The Origin and Character of Israelite Ancestor Cults and Necromancy.* D.Phil. Thesis, Department of Theology, University of Oxford, Oxford: University of Oxford.

Schwemer, Daniel. 2008. "The Storm-Gods of the Ancient Near East: Summary, Synthesis, Recent Studies (Part 1)." *Journal of Ancient Near Eastern Religions* 7 (2): 121–168.

—. 2008. "The Storm-Gods of the Ancient Near East: Summary, Synthesis, Recent Studies (Part 2)." *Journal of Ancient Near Eastern Religions* 8 (1): 1–44.

Silver, Minna. 2014. "Equid Burials in Archaeological Contexts in the

Amorite, Hurrian, and Hyksos Cultural Intercourse." *Aram* 26 (1 & 2): 335-355.

—. 2014. "The Earliest State Formation of the Amorites: Archaeological Perspectives from Jebel Bishri." *Aram* 26 (1 & 2): 243–267.

Sjoberg, A.W. 1976. "In-nin Sa-gur-ra: A Hymn to the Goddess Inanna." *Zeitschrift fur Assyriologie* 65 (2): 161–253.

Solecki, Rose, Peter M. M. G. Akkermans, Anagnostis Agelarakis, Christopher Meiklejohn, and Philip E.L. Smith. 1992. "Artificial cranial deformation in the Proto-neolithic and Neolithic Near East and its possible origin: Evidence from four sites." *Paléorient* 18 (2): 83–97.

Sullivan, Rob. 2015. *Egyptian War Correspondents and the Biblical Giants.* February 27. Accessed December 17, 2016. http://www.biblearchaeology.org/post/2015/02/27/Egyptian-War-Correspondents-and-the-Biblical-Giants.aspx.

Suriano, Matthew J. 2009. "Dynasty Building at Ugarit: The Ritual and Political Context of KTU 1.161." *Aula Orientalis* 27: 105–123.

Teixidor, Javier. 1976. "The Phoenician Inscriptions of the Cesnola Collection." *Metropolitan Museum Journal* 11: 55–70.

Tharoor, Ishaan. 2014. *Why Turkey's President Wants to Revive the Language of the Ottoman Empire.* December 12. Accessed December 28, 2016. https://www.washingtonpost.com/news/worldviews/wp/2014/12/12/why-turkeys-president-wants-to-revive-the-language-of-the-ottoman-empire/.

Toorn, K. van der, and P.W. van der Horst. 1990. "Nimrod Before and After the Bible." *Harvard Theological Review* 83: 1–29.

Toorn, K. van der, Bob Becking, and Pieter Willem van der Horst. 1999. *Dictionary of Deities and Demons in the Bible.* Leiden: Brill.

Veijola, Timo. 2006. "King Og's Iron Bed (Deut. 3:11): Once Again." *Studies in the Hebrew Bible, Qumran, and the Septuagint* 101: 60–76.

Vidal, Jordi. 2006. "The Origins ofthe Last Ugaritic Dynasty." *Altorientalische Forschungen* 33 (1): 168–175.

Washington's Blog. 2015. *Newly-Declassified U.S. Government*

Documents: The West Supported the Creation of ISIS. May 24. Accessed December 28, 2016. http://www.washingtonsblog. com/2015/05/newly-declassified-u-s-government-documents-the- west-supported-the-creation-of-isis.html.

White, Chris. 2013. *Daniel: A Commentary.* CWM Publishing.

—. 2014. *False Christ: Will the Antichrist Claim to Be the Jewish Messiah?* CWM Publishing.

—. 2013. *Mystery Babylon—When Jerusalem Embraces the Antichrist: An Exposition of Revelation 18 and 19.* CWM Publishing.

—. 2015. *The Islamic Antichrist Debunked: A Comprehensive Critique of the Muslim Antichrist Theory.* CWM Publishing.

Whiting, Robert M. 1995. "Amorite Tribes and Nations of Second-Millennium Western Asia." *Civilizations of the Ancient Near East* 2: 1231–1242.

Wolters, Al. 1995. "Belshazzar's Feast and the Cult of the Moon God Sîn." *Bulletin for Biblical Research* 5: 199–206.

Young, William. 2015. *ISIS Aims to Occupy Mecca.* January 17. Accessed December 28, 2016. http://www.newsweek.com/ islamic-state-aims-occupy-mecca-300205.

NOTES

1. Black, J. A., Cunningham, G., Fluckiger-Hawker, E., Robson, E., and Zólyomi, G. "The Sumerian King List: Translation," *The Electronic Text Corpus of Sumerian Literature* (http://etcsl.orinst. ox.ac.uk/section2/tr211.htm), retrieved 12/24/16.

2. Coon, C. S. "The Eridu Crania: A Preliminary Report," *Sumer* 5, 1949, p. 103.

3. Solecki, Rose; Akkermans, Peter M. M. G.; Agelarakis. Anagnostis; Meiklejohn, Christopher; Smith, Philip E. L. "Artificial cranial deformation in the Proto-neolithic and Neolithic Near East and its possible origin: Evidence from four sites," *Paléorient*, 1992, vol. 18, no. 2, pp. 83–97.

4. Lipinski, Edward. "El's Abode: Mythological Traditions Related to Mount Hermon and to the Mountains of Armenia," *Orientalia Lovaniensa Periodica* 2, 1971, p. 19.

5. Heiser, Dr. Michael S. "The Nephilim," *Sitchin Iis Wrong.com* (http://www.sitchiniswrong.com/nephilim/nephilim.htm), retrieved 12/16/16.

6. Lipinski, op. cit., p. 69.

7. Black, J. A., Cunningham, G., Fluckiger-Hawker, E, Robson, E., and Zólyomi, G. "Enmerkar and the Lord of Aratta," *The Electronic Text Corpus of Sumerian Literature* (http://etcsl.orinst.ox.ac.uk/cgi-bin/etcsl.cgi?text=t.1.8.2.3#), retrieved 12/17/16.

8. Black, J. A., Cunningham, G., Fluckiger-Hawker, E., Robson, E., and Zólyomi, G. "A *cir-namcub* to Inana (Inana I)," *The Electronic Text Corpus of Sumerian Literature* (http://etcsl.orinst.ox.ac.uk/cgi-bin/etcsl.cgi?text=t.4.07.9&charenc=j#), retrieved 12/17/16.

9. Sjoberg, A. W. "In-nin Sa-gur-ra: A Hymn to the Goddess Inanna," *Zeitschrift fur Assyriologie* 65, no. 2 (1976): p. 225.

10. *Enmerkar and the Lord of Aratta*, op. cit.

11. Ibid.

12. Kramer, Samuel N., "The Babel of Tongues: A Sumerian Version," *Journal of the American Oriental Society* 88:108–11, 1968.

13. Safar, Fu'ād; Lloyd, Seton; Muṣṭafá, Muḥammad 'Alī; Mu'assasah al-'Āmmah lil-Āthār wa-al-Turāth. *Eridu*. Republic of Iraq, Ministry of Culture and Information, State Organization of Antiquites and Heritage, Baghdad, 1981.

14. Gordon, Cyrus H. "Abraham and the Merchants of Ura," *Journal of Near Eastern Studies* 17:1 (1958), pp. 28–31.

15. Rendsburg, Dr. Gary. "YHWH's War Against the Egyptian Sun-God Ra: Reading the Plagues of Locust, Darkness, and Firstborn in their Ancient Egyptian Context," *TheTorah.com* (http://thetorah.com/yhwhs-war-against-the-egyptian-sun-god-ra/), retrieved 12/24/16.

16. Gardiner, Alan H. "Late-Egyptian Stories," *Bibliotheca Aegyptiaca* 1, Bruxelles, 1932.

17. See Exodus 29:13 and 22 and Leviticus 3:4, 10, and 15, for example.

18. Teixidor, Javier. "The Phoenician Inscriptions of the Cesnola Collection," *Metropolitan Museum Journal* 11 (1976), p. 65.

19. Missler, Dr. Chuck. "The Military Book of the Old Testament: The

Conquering Hero—Joshua," *Koinonia House* (http://www.khouse.org/articles/2001/319/), March 19, 2001, retrieved 12/17/16.

20. Livingston, Dr. David. "The Fall of the Moon City," http://davelivingston.com/mooncity.htm, retrieved 12/17/16.

21. Yes, Rosh Hashanah is in the fall. The Jewish calendar essentially has two new years—Nisan is the first month of the year, but the year numbers change in the month of Tishri.

22. Pritchard, James B., editor. 1958. *The Ancient Near East: Volume I, An Anthology of Texts and Pictures*, p. 225. Princeton, New Jersey: Princeton University Press.

23. Sullivan, Rob. "Egyptian War Correspondents and the Biblical Giants," Associates for Biblical Research, February 27, 2015 (http://www.biblearchaeology.org/post/2015/02/27/Egyptian-War-Correspondents-and-the-Biblical-Giants.aspx), retrieved 12/17/16.

24. Chiera, Edward & Kramer, Samuel Noah & University of Pennsylvania. University Museum. Babylonian Section. (1934). *Sumerian Epics and Myths*, Chicago, IL: The University of Chicago Press, Plate nos. 58 and 112.

25. Bodi, Daniel. "Is There a Connection Between the Amorites and the Arameans?", *Aram* 26:1 & 2 (2014), p. 385.

26. Pritchard, James B., editor. 1958. *The Ancient Near East: Volume I, An Anthology of Texts and Pictures*, p. 616. Princeton, New Jersey: Princeton University Press.

27. Veijola, Timo. "King Og's Iron Bed (Deut. 3:11): Once Again," *Studies in the Hebrew Bible, Qumran, and the Septuagint* (ed. Peter W. Flint et al.; VTSup 101; Leiden/Boston: Brill, 2003) p. 63.

28. Suriano, Matthew J. "Dynasty Building at Ugarit: The Ritual and Political Context of KTU 1.161," *Aula Orientalis* 27 (2009), p. 107.

29. Annus, Amar. "Are There Greek Rephaim? On the Etymology of Greek *Meropes* and *Titanes*," *Ugarit-Forschungen* 31 (1999), pp. 13–30.

30. Pritchard, op. cit., p. 616.

31. Annus, Amar. "The Suteans and Sons of Seth," *The Melammu Project: The Heritage of Mesopotamia and the Ancient Near East* (http://www.aakkl.helsinki.fi/melammu/database/gen_html/a0001500.php), retrieved 12/20/16.

32. Becking, Bob, "Sisera," *Dictionary of Deities and Demons in the Bible* (Leiden/Boston: Brill, 1999), p. 784.

33. For reference, an eighteen-wheel flatbed trailer carries a maximum legal load of about twenty-two tons—so load forty eighteen-wheelers to the limit and you've got the weight of just one of those massive blocks.

34. There is a Greek myth of Lycaon, king of Arcadia, who tested the omniscience of Zeus by serving him the roasted flesh of his son. But Zeus punished the king by turning him into a wolf, indicating that such behavior wasn't normal or accepted.

35. Cross, Frank Moore. "The Abode of 'El," *Canaanite Myth and Hebrew Epic: Essays in the History of the Religion of Israel,* Harvard University Press (1997), p. 36.

36. Bodi, op. cit., p.409.

37. Mello, Alex and Knights, Michael. "Losing Mosul, Regenerating Diyala: How the Islamic State Could Exploit Iraq's Sectarian Tinderbox," *Combating Terrorism Center,* October 25, 2016 (https://www.ctc.usma.edu/posts/losing-mosul-regenerating-in-diyala-how-the-islamic-state-could-exploit-iraqs-sectarian-tinderbox), retrieved 12/22/16.

38. Gadd, C. J. "The Harran Inscriptions of Nabonidus," *Extrait d'Anatolian Studies,* VIII (1958), pp. 35- 92. In: *Syria.* Tome 37 fascicule 1–2, 1960. pp. 194–196.

39. But thanks to Joel Richardson for the inspiration. Maybe as a novel…

40. "Major Religions of the World Ranked by Number of Adherents," *Adherents.com* (http://www.adherents.com/Religions_By_Adherents.html), retrieved 2/28/16.

41. "Global Christianity—A Report on the Size and Distribution of the World's Christian Population," Pew Research Center, December 19, 2011 (http://www.pewforum.org/2011/12/19/global-christianity-exec/), retrieved 12/22/16.

42. "Barna Survey Examines Changes in Worldview Among Christians over the Past 13 Years," Barna Group, March 9, 2009 (https://barna.org/barna-update/transformation/252-barna-survey-examines-changes-in-worldview-among-christians-over-the-past-13-years#.VtMY3hj5hE4), retrieved 12/22/16.

43. For the purposes of the Barna Group's 2009 survey, a biblical worldview was defined as accepting all of the following points of doctrine: Believing that absolute moral truth exists; the Bible is totally accurate in all of the principles it teaches; Satan is considered to be a real being or force, not merely symbolic; a person cannot earn his or her way into heaven by trying to be good or do good works; Jesus Christ lived a sinless life on earth; and God is the all-knowing, all-powerful creator of the world who still rules the universe today.

44. Furnish, Dr. Timothy R. "Mahdism (and Sectarianism and Superstition) Rises in the Islamic World," *History News Network*, August 13, 2012 (http://historynewsnetwork.org/article/147714), retrieved 12/28/16.

45. Furnish, 2012.

46. Byman, Daniel L. and Williams, Jennifer R. "ISIS vs. Al Qaeda: Jihadism's Global Civil War." February 24, 2015 (http://www.brookings.edu/research/articles/2015/02/24-byman-williams-isis-war-with-al-qaeda), retrieved 12/28/16.

47. Young, William. "ISIS Aims to Occupy Mecca," *Newsweek*, January 17, 2015 (http://www.newsweek.com/islamic-state-aims-occupy-mecca-300205), retrieved 12/28/16.

48. Tharoor, Ishaan. "Why Turkey's President Wants to Revive the Language of the Ottoman Empire," *Washington Post*, December

12, 2014 (https://www.washingtonpost.com/news/worldviews/
wp/2014/12/12/why-turkeys-president-wants-to-revive-the-
language-of-the-ottoman-empire/).

49. "Newly-Declassified U.S. Government Documents: The West
Supported the Creation of ISIS," *Washington's Blog*, May 24, 2015
(http://www.washingtonsblog.com/2015/05/newly-declassified-u-
s-government-documents-the-west-supported-the-creation-of-isis.
html).

50. Furnish, Dr. Timothy R. "A Western View on Iran's WMD Goal:
Nuclearizing the Eschaton, or Pre-Stocking the Mahdi's Arsenal?,"
Institute for Near East & Gulf Military Analysis, January 2011, p. 4.

51. Maghen, Ze'ev. "Occultation in Perpetuum: Shi`ite Messianism
and the Policies of the Islamic Republic," *Middle East Journal*, Vol.
62, No. 2 (Spring 2008), p. 237 (cited by Furnish).

52. Lee, Matthew. "Rice: Now is Time for Palestinian
State," *Washington Post*, October 15, 2007 (http://www.
washingtonpost.com/wp-dyn/content/article/2007/10/15/
AR2007101500703_pf.html).

53. Jerusalem Old City Initiative, http://www.cips-cepi.ca/event/
jerusalem-old-city-initiative/, retrieved 12/12/15.

54. Klein, Aaron. "International mandate to control sections of
Israel's capital," *WND*, December 15, 2013 (http://www.wnd.
com/2013/12/u-s-plan-gives-jerusalem-holy-sites-to-vatican/).

55. Blau, Uri. "Netanyahu Allies Donated to Groups Pushing for Third
Temple," *Haaretz*, December 9, 2015 (http://www.haaretz.com/
settlementdollars/1.690821), retrieved 12/27/16.

56. Berkowitz, Adam Eliyahu. "Prominent Rabbis Sternbuch, Amar
Hint That the Messiah is 'Just Around the Corner,'" *Breaking
Israel News*, December 9, 2015 (http://www.breakingisraelnews.
com/55777/turkeysyria-conflict-unfolding-prominent-rabbis-hint-
messiah-around-corner-jewish-world/), retrieved 12/28/16.

57. Adler, Dr. Rivkah Lambert. "18th Century Jewish Mystics
Predicted Future Conflict Between Russia, Turkey Ripe with

Messianic Implications," *Breaking Israel News*, November 29, 2015 (http://www.breakingisraelnews.com/54943/200-years-ago-war-between-turkey-russia-prophesied-sign-redemption-jewish-world/), retrieved 12/28/16.

58. Ostroukh, Andrey. "Russia Suggests Turkey Planned Downing of Warplane," *The Wall Street Journal*, December 23, 2015 (http://www.wsj.com/articles/russia-suggests-turkey-planned-downing-of-warplane-1450899609), retrieved 12/18/16.

59. Berkowitz, December 9, 2015, retrieved 12/18/16.

60. Berkowitz, Adam Eliyahu, "Torah Scroll Being Written to Present to Messiah Upon His Arrival," *Breaking Israel News*, December 15, 2015 (http://www.breakingisraelnews.com/56244/rabbi-sets-out-mission-write-torah-scroll-present-messiah-upon-arrival-jewish-world/), retrieved 12/18/16.

61. Adler, Dr. Rivkah Lambert, "Rabbi Kaduri 'Jesus as Messiah' Claim Proven as False," *Breaking Israel News*, June 17, 2015 (http://web.archive.org/save/http://www.breakingisraelnews.com/43554/rabbi-kaduri-jesus-as-messiah-claim-discredited-as-false-jewish-world/), retrieved 12/18/16.